Adhesion of Microorganisms to Surfaces

Special Publications of the Society for General Microbiology

PUBLICATIONS OFFICER: A. G. CALLELY

1. Coryneform Bacteria, eds. I. J. Bousfield & A. G. Callely
2. Adhesion of Microorganisms to Surfaces,
   eds. D. C. Ellwood, J. Melling & P. Rutter
3. Microbial Polysaccharides and Polysaccharases,
   eds. R. C. W. Berkeley, G. W. Gooday & D. C. Ellwood

Adhesion of Microorganisms to Surfaces

Edited by

D. C. ELLWOOD

and

J. MELLING

Microbiological Research Establishment
Porton Down
Salisbury, U.K.

P. RUTTER

Unilever Limited
Isleworth, U.K.

1979

Published for the
Society for General Microbiology
by
ACADEMIC PRESS
London New York San Francisco
A Subsidiary of Harcourt Brace Jovanovich, Publishers

ACADEMIC PRESS INC. (LONDON) LTD
24/28 Oval Road,
London NW1

United States Edition published by
ACADEMIC PRESS INC.
111 Fifth Avenue
New York, New York 10003

British Library Cataloguing in Publication Data
Adhesion of microorganisms to surfaces.
1. Micro-organisms – Physiology – Congresses
2. Adhesion – Congresses
I. Ellwood, D C
II. Melling, J III. Rutter, P R
IV. Society for General Microbiology. *Cell Surfaces and Membranes Group* V. Society for General Microbiology. *Ecology Group* VI. Society of Chemical Industry. *Microbiology, Fermentation and Enzyme Technology Group*
576'.11'8 QR84 79–50309
ISBN 0–12–236650–6

Printed in Great Britain by Whitstable Litho Ltd, Whitstable, Kent

LIST OF CONTRIBUTORS

J.P. Arbuthnott, *Department of Microbiology, Trinity College, Dublin 2, Ireland.*

S.G. Ash, *Shell Research Limited, Shell Biosciences Laboratory, Sittingbourne Research Centre, Sittingbourne, Kent ME9 8AG.*

R.G. Burns, *Biological Laboratory, University of Kent, Canterbury, Kent CT2 7NJ.*

A.S.G. Curtis, *Department of Cell Biology, University of Glasgow, Glasgow G12 8QQ.*

D.C. Ellwood, *Microbiological Research Establishment, Porton Down, Salisbury, Wiltshire SP4 0JG.*

Madilyn Fletcher, *Department of Environmental Sciences, Coventry, Warwickshire CF4 7AL.*

N.E. Jessup, *Unilever Research, Port Sunlight, Wirral, Merseyside, L62 4XN.*

A. Lips, *Unilever Research, Port Sunlight, Wirral, Merseyside L62 4XN.*

J. Melling, *Microbiological Research Establishment, Porton Down, Salisbury, Wiltshire SP4 0JG.*

H.J. Rogers, *National Institute for Medical Research, Mill Hill, London NW1 7AA.*

P. Rutter, *Unilever Research, Isleworth Laboratory, 455 London Road, Isleworth, Middlesex TW7 5AB.*

C.J. Smyth, *Department of Bacteriology and Epizootology, Swedish University of Agricultural Sciences, College of Veterinary Medicine, Biomedical Centre, Uppsala, Sweden.*

PREFACE

Early in 1977 the amount of current interest in the adhesion of microorganisms to surfaces indicated that a meeting on this topic would be profitable, particularly if it served to draw together for discussion biologists from one end of the spectrum of workers active in this field and physical chemists from the other. It was decided that one way of attempting to achieve this was to organise a joint meeting between the Cell Surfaces and Membranes Group and the Ecology Group of the Society for General Microbiology and the Microbiology, Fermentation and Enzyme Technology Group of the Society of Chemical Industry. In an unusually short space of time the arrangements were made and some 150 participants met on 1st December 1977 in the lecture theatre of the Society for Chemical Industry, filling it to capacity. This book arose out of this meeting, all the invited speakers contributing chapters. In addition there are summarising comments by Professor A.S.G. Curtis, and an introduction by the Editors one of whom, Professor D.C. Ellwood, took the chair at the meeting.

We would like to thank the Staff of the two Societies involved for their assistance in arranging this meeting at such short notice and also Unilever and ICI for contributions towards the expenses incurred. It is hoped that a second meeting on this topic will be held in the autumn of 1980.

R.C.W. Berkeley, *Convener, Microbial Surfaces and Membranes Group of the Society for General Microbiology*

J.M. Lynch, *Convener, Ecology Group of the Society for General Microbiology.*

J. Melling, *Secretary, Microbiology Fermentation and Enzyme Technology Group of the Society for Chemical Industry.*

ACKNOWLEDGEMENTS

On behalf of the Society and also myself, I would like to thank everyone concerned in the preparation of this, the second in the series of "Special Publications of the Society for General Microbiology"; in particular I would like to mention the contributors for the chapters they have written, the editors for the time and energy that they have freely given in looking after the scientific side of the contents of this book, Mrs. M. Adams of the Cardiff University Industry Centre who typed and set out all these pages, and to those at Academic Press, especially Mr. A. Watkinson and Mrs. J. Sibley for their advice about presentation, lay-out and general assistance.

The three editors should not be held responsible for any typographical errors that the reader might find; these are my responsibility as I had the task of reading through all the typescripts and checking the final proofs.

A.G. Callely
(Publications Officer:
Society for General Microbiology).
Department of Microbiology,
University College,
Cardiff, Wales.

CONTENTS

INTRODUCTION

D.C. ELLWOOD, J. MELLING and P. RUTTER*

*Microbiologial Research Establishment, Porton Down, Salisbury, Wiltshire SP4 OJG and *Unilever Research Isleworth Laboratory, 455 London Road, Isleworth, Middlesex TW7 5AB.*

THE MICROBIAL SURFACE

The process(es) by which a microorganism becomes attached to a surface, be it the surface of another living cell or some non-viable support, are complex as the succeeding chapters in this book will show. It is clear however that the nature of the microbial surface is of paramount importance in both the initial and subsequent stages of the attachment process. In this brief introduction it is our intention to outline, mainly for the benefit of readers who are not microbiologists, some salient features of the microbial surface and, in particular, to emphasise the dynamic and plastic nature of this region of the cell.

The vast majority of bacteria can be divided into two groups with respect to a simple staining reaction devised by Gram. Composition of the envelope varies markedly between the so-called Gram-positive and Gram-negative bacteria and these are worth considering separately.

A major component (up to 80%) of the wall of Gram-positive bacteria is the peptidoglycan. This consists essentially of a polysaccharide chain with alternate residues of *N*-acetylglucosamine and *N*-acetyl muramic acid linked β(1-4). Most of the carboxyl groups of the *N*-acetyl muramic acid are substituted with peptide side chains which act as bridges from one glycan chain to another so forming the vast network of interlinked chains which make up the bacterial cell wall. The peptide side chains are composed of basic unit usually L-ala-D-Glu-lysine (or diaminopimelic acid)-D-ala. These side chains may be the bridge from one glycan chain to another or a further peptide may act as the bridge, as for example the pentaglycine residues in the wall of *Staphylococcus aureus*.

The peptidoglycan is fairly rigid and is maintained in close contact with the cytoplasmic membrane by the turgor pressure. However the peptidoglycan layer has an 'open' structure and is interspersed with other wall components such as lipids, and teichoic and teichuronic acids.

The teichoic acids are polyol phosphate compounds carrying sugar substituents or ester bound alanine [Baddiley, 1972]. They occur in the wall of some Gram-positive bacteria, and have been shown to be important in phage binding in *Bacillus subtilis*. Teichoic acids are also found in most Gram-positive bacteria as lipoteichoic acids (LTA), that is teichoic acids with a glycolipid end. This lipid tail anchors them in the membrane and they can extend out of the wall into the external milieu. They are apparently also secreted into the environment.

It is tempting to speculate that they may play a role in microbial interaction because of their sugar substituents giving a specific effect, or non-specifically due to their high ionic charge. Most bacterial surfaces are indeed negatively charged and as teichoic acid in walls is maximal when an organism is grown under conditions of magnesium limitation it seems reasonable to suggest that it plays a role in magnesium ion uptake in Gram-positive bacteria. It may be that the teichoic acid of one organism could be joined by a divalent cation to the teichoic acid of another organism.

In many Gram-positive bacteria the walls contain only small amounts of protein, but in Streptococci the M, T and R proteins make up an important part of the wall. At the surface of Gram-positive bacteria globular proteins or polysaccharide capsules may occur. The polysaccharides have been shown to be important antigens in streptococci, pneumococci and lactobacilli.

The innermost layer of the more complex Gram-negative wall is again peptidoglycan, but this compound accounts for a much smaller portion of the wall (about 8 - 12%) than is the case with Gram-positive bacteria and is not interspersed with other polymers. In many, but not all, Gram-negative bacteria covalently attached to the peptidoglycan are lipoproteins which extend, almost like pillars, to an outer membrane and leave a periplasmic space. The outer membrane contains lipopolysaccharide molecules (LPS) whose lipid A position is located in the hydrophobic continuum. In smooth strains of Gram-negative bacteria the hydrophilic polysaccharide moities project from the cell surface.

Surface (capsular) polysaccharides also occur in Gram-negative organisms and may well be important in cell-to-cell interactions because of the high degree of specificity that can derive from the number of structures possible in such compounds. A similar argument pertains to the polysaccharide portion of the LPS. Indeed the LPS plays a vital role in some surface-associated properties of Gram-negative bacteria such as antigenic classification and phage attachment.

Proteins form a major part of the Gram-negative cell wall both as pure proteins which function as enzymes, lipoproteins and capsular proteins. They are involved in transport and in the receptors for colicin, phages and pili and like the LPS are subject to phenotypic and genotypic variation.

Surface appendages include flagella, common pili (fimbriae) and sex pili. Flagella are involved in locomotion, but have no certain role in the adhesion process. Common pili certainly exhibit some sugar specificity and have been shown to function as haemagglutinins (proteins with specific blood group agglutinating activity). The sex pili of *Escherichia coli* are made up of a protein of sub-unit molecular weight 12,000 containing one glucose and two phosphate residues per sub-unit, and are rich in amino acids with non-polar side chains. These pili are essential for conjugation. It is not clear how the initial interaction between the donor and recipient takes place but it may be hydrophobic in nature as the pili have large regions of hydrophobic amino acids. Nor is it clear how the DNA is transferred. Sex pili are several times longer than common pili.

There are mating substances in yeasts, protozoa and *Chlamydomonas*, but little is known of the chemistry of the interactions except that glycoproteins may be involved and the specificity of the sugar residues may play a role in recognition.

DYNAMIC NATURE OF THE MICROBIAL SURFACE

That wall-turnover occurs in bacteria is now well established. It was first observed in 1964 for the peptidoglycan of *Bacillus megaterium* and since then has been shown with respect to the anionic wall polymers of *Bacillus subtilis* and the peptidoglycan of several Gram-positive bacteria, but not Gram-negative organisms. However, secretion of LPS has been established which may be analagous to secretion of lipoteichoic acid (LTA) by Gram-positive bacteria.

Thus turnover of walls in bacteria may be compared with that in membranes of eukaryotic cells. Interaction between eukaryotic cells has been related to the glycosyl-transferase enzymes of one cell binding to the unfinished glycopeptide of another cell leading to cell adhesion. As cell wall turnover in bacteria occurs, it is possible, though speculative, that the enzymes responsible for peptidoglycan, teichoic acid and LPS synthesis could bind to the corresponding unfinished substrate in adjacent bacteria and this mechanism could be a basis of the (initial) interaction. Many bacteria produce substances which bind to sugars of specific stereochemistry and several bacterial haemagglutinins have been recognised. These substances may play a role in microbial interactions, as in, for example, the binding of pathogens to specific sites in the body. Plants also produce these sugar-specific proteins (lectins) and they may play a role in binding the specific bacteria associated with them, for example legumes and Rhizobia.

Along with the observation that wall turnover occurs it has also been recognised that the bacterial cell wall is subject to very considerable phenotypic variation in composition and properties. Thus bacteria which have undergone transition from one environment to another may possess 'old wall' and 'new wall'; this was well illustrated by the retention of phage-binding ability only at the poles of *Bacillus subtilis* following a transition from potassium-limited to phosphorus-limited growth. Such 'differential' wall replacement can obviously produce cells with a spectrum of adhesive capabilities which may be of value in natural environments and could be involved in polar attachment.

In nature it is usual that essential nutrients for microorganisms are in short supply and thus attachment of organisms to surfaces may position them in a nutrient stream and additionally, nutrient concentration at such surfaces may also occur. An organism so positioned would avoid expenditure of energy in moving along a nutrient concentration gradient.

Wall turnover may also be responsible for the release of various enzymes which would in effect allow an organism to "survey" its environment as the products of such enzyme activity became available to it and possibly result in further phenotypic modification.

Thus it is possible to conclude, and this will be amply illustrated in the following chapters, that the ability to adhere to surfaces provides an important survival mechanism for microorganisms and one which must be investigated in detail using techniques available to both bacteriologists and physical chemists if the wide variety of microbial interactions in nature is to be fully understood.

COLLOIDAL ASPECTS OF BACTERIAL ADHESION

A. LIPS and N.E. JESSUP

Unilever Research, Port Sunlight, Wirral, Merseyside, L62 4XN.

INTRODUCTION

The adhesion of particles, both living and inert, is of interest to scientists and technologists from a wide range of disciplines and, in the main, such phenomena have been studied from a colloid science viewpoint. In recent years, hydrodynamicists and engineers have paid considerable attention to particulate deposition in various externally applied fields, and in many deposition situations [Spielman and Friedlander, 1974] including porous substrates (such as, for example, soil and filters). The fundamental description of the adhesion of biological systems is proving to be particularly challenging, and to this end there have been significant attempts at bringing together the physical and biosciences.

It is our intention in this article to outline the principal physical factors that may determine the strength of the adhesion of a particle to an interface, and to have special regard for these factors in the context of living organisms depositing on substrates. This is not so much intended as a review of bacterial deposition - excellent articles exist on this and related topics (see, for example, Marshall, 1976; Weiss and Harlos, 1972) - but rather as a broad view of the usefulness of physical concepts to biological applications.

GENERAL REMARKS ON FORCES

The interaction energy $V(\ell)$ of a particle and a surface is the energy required to bring the particle from the bulk aqueous phase to a distance ℓ from the surface. For adhesion to occur $V(\ell)$ must be negative for at least some values of ℓ. Frequently what is required is not only $V(\ell)$ but also the force, $(-\partial V/\partial \ell)$ acting between the particle and the substrate.

If the particles are large or, more strictly, if Brownian thermal forces are small compared with the forces of attachment, a deterministic approach which is based on the sole consideration of the balance of forces acting on individual attached particles, is possible. On the other hand, if the particles are small and interact weakly with the substrate, a probabilistic approach is more appropriate; this requires a statistical mechanical treatment. In its simplest form, such a treatment assumes that the probability of finding a particle at a distance ℓ from the substrate is given by the Boltzmann factor $\exp[-V(\ell)/kT]$ where k is the Boltzmann constant and T the absolute temperature. The distinction between a deterministic and probabilistic treatment can be of importance in the analysis of particle attachment measurements [Dahneke, 1975].

The forces to be considered can be divided into three main types. There are the conservative forces which are consequences of the mutual proximity of media, arising solely by virtue of the positions of media relative to one another. In this category fall, for example, electrodynamic and electrostatic forces. Other forces are those arising from the movement of particles in liquids or in air (hydrodynamic and aerodynamic forces). Finally there are externally generated forces such as electric and magnetic fields, hydrodynamic shear and gravitational forces.

It should be noted that hydrodynamic forces are present in the absence of external shear fields. Their effect then is to mediate dynamic factors such as particle diffusion or the rate at which a system approaches a new equilibrium position (kinetics of deposition, particle aggregation). In the absence of external energy input, however, hydrodynamic forces do not affect particle-particle or particle-substrate equilibria; these are expected to be controlled solely by conservative interactions. It has to be recognised though that, in many if not most biological situations, external forces operate.

The adhesion of microorganisms cannot always be discussed satisfactorily on the basis of only conservative forces such as electrodynamic or electrostatic forces. Before entering into a detailed description of the various forces it should be remembered that living systems can undergo significant morphological and biochemical changes on contacting substrates. The forces are expected therefore to be time dependent. Reference will be made to this possibility in the discussions on particle deformation and polymer mediated interactions. In the following we shall confine our attention to aqueous systems and, in particular, to interactions of particles with solid/water interfaces. Nevertheless, the concepts which will be described extend in application to liquid/water and other interfaces.

DISPERSION FORCES

Molecules in proximity experience attractive dispersion interactions, and the dispersion energy (also called the van der Waals or electrodynamic energy) of a pair of non-polar molecules at moderate separation varies as R^{-6} where R is their separation. The attractive energy of two particles in proximity, each composed of a large number of molecules, was assumed by Hamaker [1937] to be given by the sum of all possible interactions between pairs of molecules on different particles. The resultant energy, which is usually attractive, between flat media is predicted to vary as $1/\ell^2$ where ℓ is the separation. The attractive force arises from the coupling of electromagnetic fluctuations and encompasses interactions over the entire frequency spectrum of electromagnetic radiation.

In the Hamaker theory, which confines attention to a narrow spectral range, the characteristic constant describing the inverse square dependence on ℓ is material dependent and is referred to as the Hamaker constant. A number of prescriptions have been given for its estimation for any combination of three media in contact. The theory, however, is recognised to be inadequate and not until recently when dramatic progress was achieved in the theory of van der Waals dispersion forces has the calculation of dispersion forces become reliable.

This progress can be traced back to Liftshitz [1955,1956], who has drawn together earlier theories of Keesom, Debye, London, Casimir and Hamaker, and to Dzyaloshinskii, Lifshitz and Pitaevskii [1961] who in a more general way have given a framework for calculating dispersion interactions for any combination of dielectric media from measurements of their spectral properties. These theories were difficult to apply and their recent exploitation in the colloid field awaited semi-classical approaches by van Kampen, Langbein, Ninham and Parsegian and others (see Richmond, 1975, for a detailed review and references). From these emerged relatively simple prescriptions for calculating van der Waals interactions which are also addressed to the difficulties resulting from an incomplete knowledge of the dielectric properties of materials over the entire electromagnetic spectrum.

Exploitation of the recent developments was sought especially in the biological area (see, for example, Parsegian and Gingell [1973] and a review by Parsegian [1973]). This directed attention to the importance of low frequency fluctuations to the dispersion interactions. Unlike the high frequency terms (visible and uv), the low frequency contributions to the dispersion energy are strongly temperature dependent and, moreover, can be shielded by electrolyte. As will be seen, the latter prediction has important implications for interactions involving biological systems.

It is usual to define the dispersion free energy $V_A(\ell)$ between flat media separated by a distance ℓ as

$$V_A(\ell) = -\frac{A(\ell,T)}{12\pi\ell^2} \qquad (1)$$

where $A(\ell,T)$ is now more appropriately referred to as a 'Hamaker function' in contrast to the earlier Hamaker constant. (The original Hamaker theory was concerned essentially with the largely temperature independent ultraviolet contribution to the dispersion energy).

Strictly $A(\ell,T)$ represents the summation of contributions from fluctuations over the entire frequency range. For low density, dielectric materials, for example, hydrocarbon layers or cell walls, interacting across water the so-called 'zero frequency' contribution to $A(\ell,T)$ is approx. $0.75(kT)^{-1}$ [Parsegian and Ninham, 1970]. Predictions of the sum of all other contributions (typically 60% from the infra red and 40% from the visible and ultra-violet) lie in the range, 4×10^{-21}J to 8×10^{-21}J; this corresponds to $\sim$ 1 to 2 kT at 300K.

The low frequency contribution is thus potentially of importance in the biological context. However, because of its sensitivity to electrolyte concentration its importance is often diminished in practical applications. Recent theory predicts this term to be screened according to $e^{-2\kappa\ell}(\kappa\ell)^{-1}$ where $1/\kappa$ is the characteristic Debye-Huckel screening length [Richmond, 1975] ($\kappa^2 = 2N_A Z^2 Ce^2/\varepsilon kT$, C being the concentration of electrolyte, ε the dielectric constant of the medium and e the proton charge. N_A is Avogadro's Number and Z the counter-ion valency). In physiological saline, $1/\kappa \simeq 1$ nm; for separations $\ell > 1$ nm the low frequency term makes therefore no effective contribution.

It is not realistic in the context of microorganisms to employ the present theories to separations below 2 nm (see later). Nevertheless, successful predictions can be made at larger separations. Because the zero frequency contribution is then of no consequence, the theoretical prediction of van der Waals forces based on simple Hamaker theory is not quite as much in error as has sometimes been implied. Nevertheless, it has to be recognised that a major contribution to the 'Hamaker constant' of hydrocarbon media interacting across water does originate from fluctuations at infra-red frequencies and not solely from those of the visible and ultraviolet regions. The dinstinction is especially important when one considers dispersion forces at long range where retardation effects are important.

Though considerably more complex than the earlier Hamaker theory, the recent treatments have given colloid scientists increased confidence in the calculation of a major contribution to interactions of materials in close proximity. Especially noteworthy are the extensive theoretical calculations of Parsegian and Gingell [1973] of dispersion forces between cells and cells on substrates. There have also been attempts at calculating effective Hamaker constants for a number of biological situations [van Oss *et al.*, 1977].

ELECTROSTATIC FORCES

As most surfaces, including those of microorganisms, bear a net electric charge a consideration of electrostatic forces is crucial to any discussion of cell-substrate interactions. Despite considerable theoretical activity in this area the detailed understanding of electrostatic forces is not as advanced as the theories of dispersion forces.

The basic problem in determining electrostatic forces is to determine the distribution of ions on isolated surfaces and then to consider how this distribution, and hence the energy, is perturbed as two surfaces approach one another. This involves the Poisson-Boltzmann equation

$$\nabla^2 \psi(\underset{\sim}{r}) = \frac{e}{\varepsilon} \sum_i n_{io}\, z_i \exp\left[\frac{-\psi(\underset{\sim}{r})\, z_i e}{kT}\right] \qquad (2)$$

where n_{io} is the bulk concentration of ions of species i and charge z_i and $\psi(\underset{\sim}{r})$ is the electric potential in the region $\underset{\sim}{r}$; e is the proton charge and ε the dielectric constant of the medium. A major problem is that equation (2) cannot be solved analytically. If $\psi(\underset{\sim}{r})e \ll kT$ then

$$\nabla^2 \psi(\underset{\sim}{r}) \simeq \frac{e}{\varepsilon} \sum_i n_{io}\, z_i \left[1 - \frac{\psi(\underset{\sim}{r}) z_i e}{kT}\right] \qquad (3)$$

which is the so-called linearised Poisson-Boltzmann equation which is solvable subject to certain assumptions. We wish to know the electrostatic interaction energy, V_R, which is the change in the free energy of the system when two surfaces are brought from infinity to the separation of interest:

$$V_R(\ell) = G_R(\ell) - G_\infty \qquad (4)$$

where G_∞ is the sum of the free energies at infinite separation.

To achieve analytic solutions of (3), two limiting assumptions are made - either that the surfaces approach at constant electrical surface potential or that the approach is at constant surface charge. For the particular case of a spherical particle of radius a interacting with an infinite planar substrate [Hogg, Healy and Fuernstenau, 1966; Wiese and Healy, 1970]:

$$V_R^{\psi}(\ell) = \pi\varepsilon a \left\{ 2\psi_1\psi_2 \ \ell n \left[\frac{\exp(\kappa\ell) + 1}{\exp(\kappa\ell) - 1} \right] + (\psi_1^2 + \psi_2^2)\ell n \left[\frac{\exp(2\kappa\ell) - 1}{\exp(2\kappa\ell)} \right] \right\} \quad (5a)$$

$$V_R^{\sigma}(\ell) = \pi\varepsilon a \left\{ 2\psi_1\psi_2 \ \ell n \left[\frac{\exp(\kappa\ell) + 1}{\exp(\kappa\ell) - 1} \right] - (\psi_1^2 + \psi_2^2)\ell n \left[\frac{\exp(2\kappa\ell) - 1}{\exp(2\kappa\ell)} \right] \right\} \quad (5b)$$

where V_R^{ψ} and V_R^{σ} are, respectively, the electrostatic interactions at constant surface potential and constant surface charge. ψ_1 and ψ_2 are the surface potentials of the planar surface and sphere at infinite separation. Electrokinetic measurements, of course, afford an approximate measure of these.

To illustrate the nature of the above expressions, some model calculations, which span physiological conditions, are shown in Figures 1-3 ($\varepsilon = 7.10^{-10}C^2m^{-1}\ J^{-1}$, $1/\kappa = 0.83$ nm, a = 500 nm, $\psi_1 = -20$mV and ψ_2 has been given values in the range ± 20 mV). Figures 2 and 3 show that at large separations $\ell \gtrsim 3/\kappa$ both expressions give similar predictions but they progressively diverge with decreasing surface separation. It is apparent that, except for the special case of $\psi_1 = \psi_2$, at sufficiently small separations the constant surface potential model predicts an attractive potential. The converse applies under conditions of constant surface charge which predicts repulsion except when $\psi_1 = -\psi_2$. Equations 5a and 5b were derived using the linearised Poisson-Boltzmann equation and so are only valid for small surface potentials ($|\psi| < 25$ mV). Bell and Peterson [1972], using a more rigorous approach, have shown that essentially the same conclusions can be reached without invoking this approximation.

Clearly then there is a difficulty in predicting electrostatic forces at short separations $\ell \lesssim \kappa^{-1}$. The models described here are limiting cases and are recognised to be inadequate. A more realistic theoretical representation must consider the approach of surfaces at constant electrochemical potential. Ninham and Parsegian [1971] have developed an approach that is especially appropriate to biological surfaces on which charge arises principally from dissociating acidic gro-

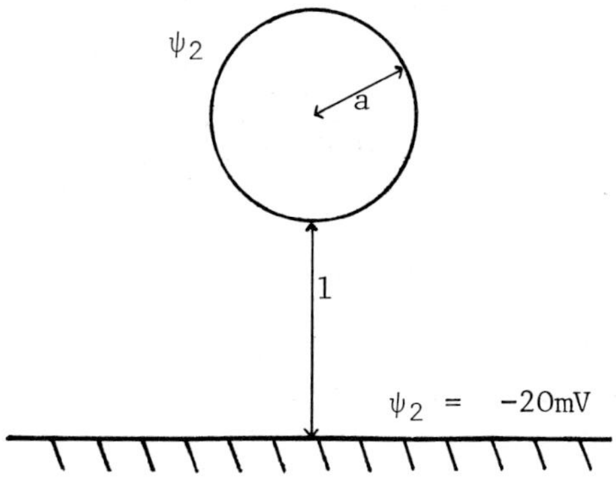

Fig. 1. Geometry for the calculation of electrostatic energies V_R^σ *and* V_R^ψ *(equations 5a and 5b) for a spherical particle, of radius* a, *interacting with a planar substrate.* ψ_1 *and* ψ_2 *are, respectively, the surface potentials of the substrate and particle at infinite separation.*

ups. These workers propose that the degree of surface ionisation when two surfaces approach is such as to maintain electrochemical equilibrium with the bulk solution.

A further refinement is to recognise that the surface charge of cells is not located within an infinitely thin shell but is spread throughout a 'fuzzy layer' several nanometers thick, and into which counterions may penetrate [Parsegian and Gingell, 1973]. These improvements render the calculation of electrostatic forces considerably more complex; they change the magnitude of the interactions but not their general shape which remains essentially exponential.

Comparatively little attention has been paid to the role of the cell interior in electrostatic phenomena. Calculations by Ohshima [1977] within the constant surface charge approximation, indicate that for $\kappa\ell > 1$ the cell interior does not significantly influence the calculated energies. At smaller separations the possibility of migration of ions within the cell can lead to reduced repulsion but caution is necessary when applying continuum electrostatic theory at small separations where, for example, the use of a macroscopic dielectric constant to characterise the aqueous medium is of dubious validity and the essentially discrete nature of the mobile charges may assume significance. Additional complications of course are that cell surfaces are not uniform and can have areas of positive charge.

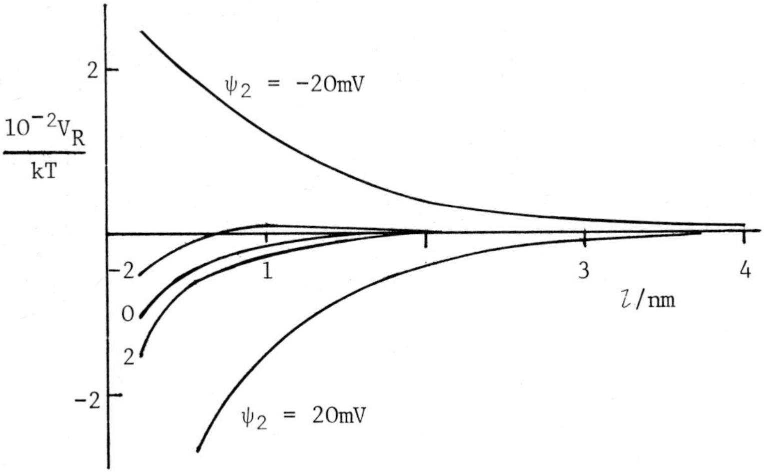

Fig. 2. Electrostatic interaction energies under conditions of constant surface potential, for different values of ψ_2.

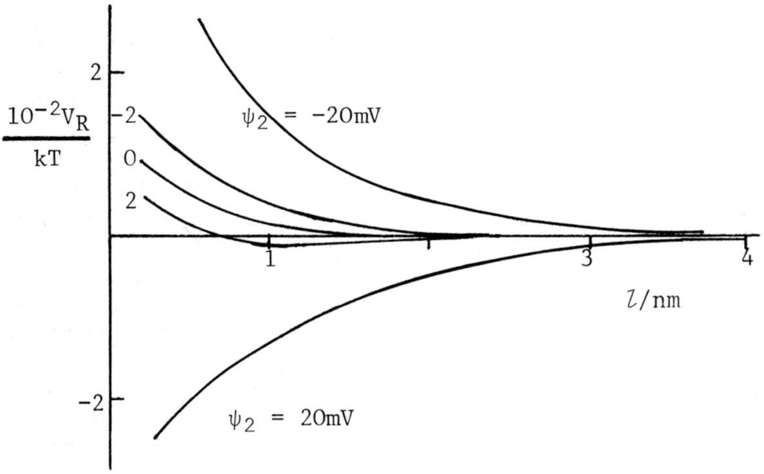

Fig. 3. Electrostatic interaction energies under conditions of constant surface charge for different values of ψ_2.

[Weiss, 1974] and more important is the possibility of specific adsorption of ions on surfaces.

Thus it can be misleading to imply that the only role of inorganic ions is to regulate the long range electrostatic interactions between the microorganism and substrate. For example, Okada *et al.* [1974] have confirmed earlier reports that cells will readily adhere to inert substrates in the absence of both serum proteins and divalent cations, while divalent cations are necessary when serum proteins are present. Furthermore, they found that Mg^{2+} is more effective than Ca^{2+} in promoting adhesion. These observations are not typical of inert colloidal systems.

DLVO THEORY

It is useful at this stage to remark on the DLVO theory which constitutes a major landmark in colloid science. It was developed by Derjaguin and Landau [1941] and Verwey and Overbeek [1948] in connection with the stability of aqueous dispersions of lyophobic particles with respect to mutual aggregation in the presence of simple electrolytes. The principles of this theory are readily applied to other systems such as the interaction of a particle with a planar substrate. The interaction energies considered are solely of the electrostatic and van der Waals type, the total energy of interaction being given by their summation:

$$V_T(\ell) = V_R(\ell) + V_A(\ell) \qquad (6)$$

Much attention has focussed on improving the theoretical representations of V_A and V_R along the lines outlined in the previous sections but in essence these have not invalidated the original expressions. We shall comment later on forces other than electrostatic and dispersion forces and concentrate at present on matters of applicability of simple DLVO theory to biological systems. Fig. 4 shows a typical energy-separation profile of the adhesion of a microorganism to a planar hydrocarbon substrate. Two regions of attraction can be discerned. That at longer range, typically 5 to 8 nm, is known as the secondary minimum and is amenable to relatively accurate theoretical prediction. The depth of the secondary minimum is typically 10^{-7} to 10^{-6} Jm^{-2}. That of the primary minimum (at much shorter range) cannot be predicted with any certainty but it is likely to be of the order of interfacial energies ($\sim 10^{-4}$ Jm^{-2}). The fact that in the region of the secondary minimum $\kappa\ell >> 1$, enables simple expressions to be used for V_A and V_R, for example, equations (1) and (5).

For typical conditions of bacterial deposition, DLVO type

Fig. 4. Typical DLVO potential energy diagram. On the left the attractive and repulsive components are shown separately. The resultant energy on the right, shows the energy maximum that separates the wide, shallow secondary minimum from the deep primary minimum.

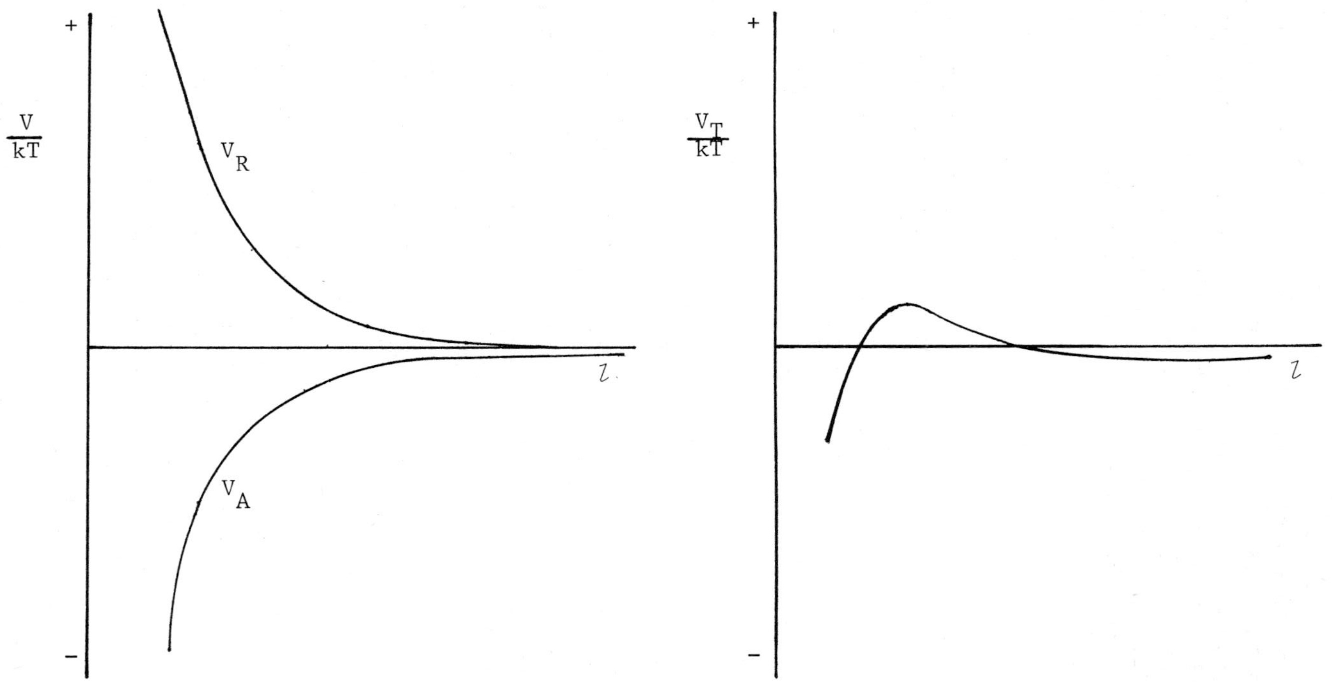

calculations reveal the following:

If the bacteria are small and their number concentration in dispersion is low the depths of secondary minima may be insufficient to give significant bacterial aggregation or deposition into such minima. The magnitude of bacterial deposition onto a flat substrate can be calculated, for known interactions using the statistical mechanical treatment, for example, of Hill [1960].

$$\frac{\Gamma}{a} = \rho\int_0^\infty \{\exp(-V_T/kT) - 1\}d\ell \qquad (7)$$

where ρ is the equilibrium number concentration of bacteria in dispersion, Γ the number adsorbed and a the substrate area.

If the bacteria are relatively large, of order 1μm, and if surface charges are low it is more reasonable to expect significant secondary minimum interactions. It can then be envisaged that small changes in surface charge or in the content of extracellular 'fuzz' material can result in changes of sufficient magnitude in the depth of the secondary minimum to explain preferences of interactions, for example, cell-cell recognition or specific aspects of deposition.

A similar comment applies to primary minima and the associated interaction barriers (Fig. 4). Having regard to the high ionic strength of physiological saline, the typical surface charges, and the expected van der Waals attraction, it is possible that in some cases the primary minimum is the major locus of interaction even though the secondary minimum is of sufficient depth to affect substantial deposition in its own right. (The interaction barriers may not always be high enough to kinetically prevent the establishment of primary minimum contacts). It is interesting to note that Gingell and Fornes [1975] have observed interactions of red blood cells with a water/solid interface consistent with secondary and primary minima adhesions.

Though the exclusive consideration of van der Waals and electrostatic forces is capable of accounting for specificity in principle (whether in primary or secondary minima), one has to bear in mind that other forces, including specific chemical binding, can operate; moreover, materials may deform on contact. Such factors are beyond the scope of DLVO theory and will be outlined in the following sections.

SOLVENT MEDIATED FORCES

It has been shown that phospholipid bilayers experience strong repulsive forces at small separations [LeNeveu *et al.*, 1977; LeNeveu, Rand and Parsegian, 1976]. These forces decay exponentially with a correlation length of approximately 0.2 nm and are stronger than can satisfactorily be accounted for in terms of dipole-dipole repulsions. Marcelja and Radic [1976] have suggested that such forces may arise as a result of modifications to the water structure at the lipid-water interface. It has long been recognised that the organisation of solvent has important implications for the calculation of interaction forces; however, until the recent definitive experiments on bilayers and the parallel theoretical developments, a quantitative understanding has been lacking. On the evidence so far, these forces are unlikely to affect predictions of secondary minima but they are considerable, if not dominant, at short range (< 3 nm). Finally it should be noted that for interactions between dissimilar surfaces, solvent structure forces can be either attractive or repulsive.

ADSORBED LAYER MEDIATED FORCES

Adsorbed layer mediated forces, especially those arising from the presence of adsorbed macromolecules are highly relevant to any discussion on surface interactions involving living systems. This area of colloid science has received considerable attention in recent years and significant advances have been made. Two properties that macromolecules uniquely possess are of special importance to our discussion: because of the possibility of multisegmental adsorption, polymers invariably display very high adsorption affinities, and because their dimensions are usually comparable with or larger than the range over which van der Waals and electrostatic forces are significant they can mediate interactions at long range.

As two surfaces carrying adsorbed polymer approach, the possibility exists of some of the polymer molecules having segments attached to both surfaces. This phenomenon is known as polymer bridging and results in increased attraction between the surfaces. The conditions favouring this process are relatively low coverage of the surfaces with polymer, macromolecules which protrude far away from the surface into the solution and which preferably have high surface mobility. In cases of irreversibly anchored polymer molecules at saturation surface coverage, the possibility of polymer bridging is greatly reduced and the mediating function of the polymer is different in character.

If the polymer is in a good solvent, as characterised by the Flory-Huggins parameter (χ) the adsorbed layers afford steric protection because the approach of surfaces requires the removal of solvent from the region between the surfaces (polymer is assumed to remain). In a good solvent, this is clearly not favoured on thermodynamic grounds. In a poor solvent on the other hand, this osmotic force manifests itself as an attraction rather than repulsion. One has to bear in mind, however, that insoluble macromolecular moieties are likely to be embedded in bilayers and are thus hidden from the aqueous environment.

Greig and Jones [1976] have considered the steric forces in cell attachment that can arise from the presence of surface glycoproteins with protruding water soluble moieties. The equation derived by Smitham, Evans and Napper [1975] was used in these calculations; this is

$$V_{ST}(\ell) = 2\left(\frac{V_s^2}{V_1}\right)\left(\frac{1}{2} - \chi\right)\left(\frac{1}{\ell} - \frac{1}{L}\right)(\nu i)^2 \qquad (8)$$

Here ν is the number of polymeric chains per unit area of contour length L, each chain having i segments of volume V_s; χ is the Flory-Huggins parameter which describes the polymer-solvent interaction and V_1 is the volume of solvent molecule. For polysaccharides in water, χ is about 0.45 which indicates that water is a good solvent and that repulsion between surfaces is to be expected.

Greig and Jones find that the steric forces can be large compared with electrostatic forces. Furthermore, they are highly dependent on the amount and density of surface saccharide, suggesting that surface coverage and not just the type of the adsorbed macromolecules can be a controlling factor in adhesion specificity. These conclusions are consistent with the well established understanding of the stability of typical inert colloids in the presence of synthetic polymers.

The situation is likely to be more complex as bridging (that is increased adsorption between two surfaces resulting in attraction) and osmotic factors (tending to repel surfaces) can operate simultaneously. At low coverage and when the surfaces are relatively far apart, the bridging contribution is more likely to be of significance; at short range, when the polymer between the surfaces is considerably compressed, the osmotic factor dominates. At present there exist theories for each contribution in isolation, those for example of Rubin [1965a, b, 1966] and DiMarzio and McCrackin [1965] for the bridging contribution of polymers which have complete freedom of rearrangement between approaching interfaces, and those, for example, of Hesselink, Vrij and Overbeek [1971] and Smitham *et al.* [1975] of the osmotic term. There is, however, no comprehensive theory dealing simultaneously with both contributions.

The ability of polymers to bridge between approaching surfaces, so strikingly manifest in the flocculation of inert colloids where extremely small amounts of polymer can give rise to major mediating effects, has important and often neglected implications for theoretical discussions on the attachment of microorganisms to substrates. Polymer mediated interactions can be of much greater range than van der Waals and electrostatic forces; only very small adsorbed amounts of high molecular weight polymer can thus induce attachment in deep minima at long range. This can occur under conditions of relatively high surface charge and low electrolyte where DLVO theory cannot support the existence of secondary minima of sufficient depth. The dependence on electrolyte of polymer induced attraction is quite different from that expected from DLVO theory and the observation of such is possibly diagnostic of effects of adsorbed layers. Similarly, differences are to be expected in the dependence of the microorganism-substrate interaction on temperature (this also applies to solvent mediated forces).

There have been attempts to take account of the presence of polysaccharide surface layers in the calculation of dispersion forces [Parsegian and Gingell, 1973]. It is assumed that these can be considered as homogeneous 'fuzz' layers of known dielectric properties. This renders relatively simple the calculation of dispersion energies. Such a procedure is undoubtedly reasonable when the separation between surfaces is large compared with the average extension of the polymer molecules on the surfaces. However, for separations less or comparable with the length of polymer chains, bridging and osmotic factors are likely to operate and indeed dominate interactions and thus render inappropriate the homogeneous fuzz model. Recently, theories have been developed which attempt to incorporate all these factors but these are of considerable complexity [Chan, Mitchell and White, 1975; Clark *et al.*, 1975].

It has to be recognised that surface layers are often polyelectrolyte in nature. Though there is qualitative understanding of the manifestations of such behaviour, it is not possible at present to predict quantitatively the relative importance of osmotic, bridging and electrostatic factors with such molecules. Theoretical progress to date has been in the main with uncharged, unifunctional polymers. Recently a treatment has been described for polyelectrolyte systems [Hesselink, 1977].

We have so far emphasised the bridging mechanism by which water soluble adsorbed molecules can reduce the free energy of interaction between approaching surfaces. Increased segmental adsorption on available surface sites is not, however, the only means by which adsorbed layers can consolidate adhesion. On the basis of equation (8) we would expect attractive interaction between polymer layers only when $\chi > 0.5$ that is, when the polymer is effectively insoluble and undergoes phase separation.

The theory underlying equation (8) is applicable only to unifunctional homogeneous polymers. Many macromolecules and especially polysaccharides clearly do not fall in this category and possess additional properties. For example, polysaccharides can be water soluble (which is suggesting $\chi < 0.5$), however, they form gels which is indicative of considerable attraction between the dissolved molecules. The nature of polysaccharide linkages has been identified in some detail and the consolidation of adhesions by the formation of such links is clearly conceivable. Indeed, there is considerable evidence in favour of such a mechanism. On kinetic grounds one might expect such a process to be relatively slow, and for it to take place it is essential that the residence time of adhered microorganisms is long compared with the time for establishment of such linkages. This requirement is more likely to be met with larger microorganisms whose attachment is in deeper secondary minima which ensures a greater stability to thermal forces.

HYDRODYNAMIC FORCES

Thus far interactions have been considered in the absence of externally applied fields. In many practical situations of bacterial depositions, however, externally imposed hydrodynamic action may be operative and its consequence on adhesion can be considerable. For a given velocity of flow of solvent past a substrate, the hydrodynamic lift force acting on attached particles is roughly proportional to the third power of the dimension of the article, and sizes > 1μm can readily be sheared off surfaces under the action of only small velocity gradients [Dahneke, 1975]. Even though secondary minimum forces can keep microorganisms attached against thermal forces, these can easily be exceeded by hydrodynamic disjoining forces. In any quantitative analysis of the attachment of particles it is important therefore to consider hydrodynamic forces.

The issue is often not so much one of a complete lack of adhesion affinity of particles for surfaces but rather one of insufficiency in the context of external disjoining forces. In mineral flotation it is known, for example, that secondary minimum forces are insufficient to keep minerals attached to rising air bubbles; contact in primary minima is necessary [Richmond, 1977]. One might speculate on the necessity in bacterial adhesion of consolidating secondary minimum contacts through the formation of polysaccharide linkages. It is the practice in bacterial deposition to distinguish between weak physical contacts (probably in secondary minima) which are affected by shearing under the action for example of a water jet, and irreversible bacterial adhesion.

IRREVERSIBLE ATTACHMENT

So far the possibility of attachment into primary minima has not been discussed in any detail, though it has been noted how difficult it is to predict their depth because of the inadequate descriptions of short range forces. Also there are usually electrostatic barriers and possibly solvent force barriers to be overcome before any deposition can proceed into a primary minimum. In the context of deposition it can be useful to employ the concept of the wettability of an interface and it has been proposed to rank the adhesion propensity of cells for substrates in terms of the hydrophobicity or hydrophilicity of the various interfaces. These concepts though undoubtedly useful can also be misleading as will be shown.

Firstly it is necessary to define what is meant by the term interfacial energy. Considering the case of two hydrocarbon layers interacting across water, the interaction energy per unit area $V_T(\ell)$ will depend on dispersion, electrostatic and possibly at short range, on solvent mediated interactions and may typically be of the form shown in Fig. 4. Now the interfacial energy $-\left[\lim_{\ell\to 0} V_T(\ell)\right]$ is the work required to de-wet unit area of the interfaces hydrocarbon/water and air/water and at the same time create unit area of hydrocarbon/water interface. This involves respectively the concepts of work of adhesion W_A and of cohesion W_C. The view presented here highlights that the interfacial energy or tension can be considered as the free energy of interaction $V_T(\ell)$ between two media across a third in the limit of zero separation.

The existing theories of forces do not enable the zero separation limit of $V_R(\ell)$ to be predicted with any certainty except possibly in cases of low energy solids (for example, polytetrafluoroethylene) wetted by non-polar liquids. The difficulties are formidable with polar liquids especially water, and on high energy surfaces. It has already been remarked how difficult it is to extrapolate the existing theories of electrostatic and dispersion interactions to short separations. Theories of the local organisation of water molecules on polar surfaces are insufficiently advanced, and primarily for this reason, the prediction of surface tension and interfacial free energy is not possible. It has to be borne in mind that the interfacial energy $-\left[\lim_{\ell\to 0} V_T(\ell)\right]$ can be orders of magnitude different from the value expected on the basis of dispersion

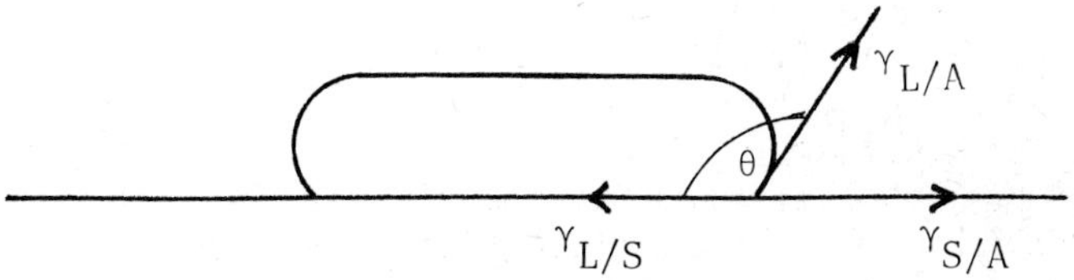

Fig. 5. Diagram to show the contact angle θ required by the Young equation.

theory; in addition its dependence on temperature can be inconsistent with that of Van der Waals forces.

The inability to predict interfacial energies should not detract from the usefulness of the concepts such as hydrophilicity, wettability and contact angle. These have been exploited in studies of bacterial phenomena; for example, measurements of the critical surface tension γ_c, following Zisman, have been performed (see Marshall, 1976 and references cited therein). Young's equation (Fig. 5) requires the following equality of interfacial energies for non-spreading liquid drops on surfaces:

$$\gamma_{S/A} = \gamma_{L/A} \cos\theta + \gamma_{L/S} \qquad (9)$$

The subscripts denote solid-air, liquid-air and liquid-solid respectively. Zisman's procedure consists of measuring the contact angle on solids for a homologous series of liquids. From a plot of liquid/air surface tension against cos Θ, the critical tension γ_c is obtained from the intercept corresponding to $\cos\theta = 1$. It is thus a measure of the highest surface tension a liquid can have and still spread over a surface. This parameter can be a useful indicator of solid gas interfacial energies especially where non-polar liquids and substrates are involved. Measurements of contact angles of water on bacteria at the air/water interface or of the partitioning of bacteria between two liquid phases can similarly provide information on relative hydrophobicities. It should be noted, however, that only $\gamma_{L/A}$ and θ in equation (9) are accessible to measurement and it is not possible therefore to determine $\gamma_{S/A}$ and $\gamma_{L/S}$ independently.

A knowledge of the hydrophobicity of bacterial surfaces can be of some value in deciding on relative deposition affinities (the strength of adhesion is expected to increase with decreasing γ_c), although a number of factors need to be taken into account. If the surface is very hydrophilic, that is spontaneously wetted by water, it may have stability conferred on it which results from the organisation of water molecules in the vicinity of the surface. This is the case, for example, with lecithin bilayers where the interaction between the layers at separations of less than about 3 nm requires the solvent organisation to be destroyed (see Solvent mediated forces). If substrate and bacteria are both hydrophilic, this stabilisation would be expected to be greater. It follows that strongly hydrophobic contacting interfaces should experience very little solvent mediated repulsion and on this basis should adhere more strongly.

In practice, however, these expectations may not always be borne out. Firstly, the wetting characteristic of a surface is essentially a manifestation of short range interactions (between the surface and air across a thin liquid film). However, long range interactions in particular those corresponding to the secondary minimum can be sufficient to give substantial deposition. The wetting behaviour of a surface is largely insensitive to these. In the main, long range interactions depend on electrolyte concentration, the magnitude of dispersion interactions and surface charge. The corresponding short range behaviour can be quite different not only in degree but also in direction. The material polytetrafluoroethylene affords a simple illustration of this. Its refractive index being close to that of water implies that its van der Waals attraction for substrates is relatively weak. It is also however a low energy surface and as such not expected to derive appreciable solvent mediated stabilisation from protective layers of organised water.

In some cases of adsorbed layer mediated interactions, the picture is similarly complex, and thus extrapolations from the wetting characteristics of surfaces can be misleading. This approach is more likely to be of value where the issue is about deposition into primary minima either because the secondary minima are of insufficient depth, or hydrodynamic, or other externally imposed conditions operate that can detach bacteria out of secondary but not primary minima.

CONTACT DEFORMATION

When solids come into contact and adhere, deformation at the points of contact is to be expected. The degree and manner of the deformation depends upon the geometry of the solids in contact, the forces between them and their elastic properties. In recent years there has been significant progress in the understanding of contact deformation and a number of theories have been advanced [Derjaguin, 1934; Derjaguin, Muller and Toporov, 1975a,b; Johnson, Kendall and Roberts, 1971;

Dahneke, 1972].

Some interesting general points can be made about contact deformation [Krupp 1967]. The interaction forces between solids in close proximity cause deformation which is more extensive in the softer, less elastic solid. The deformed solid will initially store elastic energy. If the solid is perfectly elastic this energy is permanently stored and only dissipated when the contact is broken. On the other hand the solid can be viscoelastic so that the elastic energy can be dissipated whilst the solids remain in contact. This is referred to as plastic deformation. The two types manifest themselves in major differences in the strengths of the contacts between the solids.

Although on the basis of the observed geometry of contact it is possible in principle to calculate the dispersion attraction between solids this is not the net force of adhesion if the solids are perfectly elastic as they wish to dissipate elastic energy. In fact, it has been shown that the net force of adhesion between elastic solids at a particular surface to surface separation is close to that anticipated for the case of zero deformation [Derjaguin *et al.*, 1975a,b]. In the case of plastic deformation the net force of attraction is the dispersion force calculated for the actual geometry of the deformed contact region. It follows that solids which undergo plastic deformation on contact with surfaces are more strongly attached.

Striking examples of contact deformation can be observed in the attachment of mammalian cells to substrates [Rees, Lloyd and Thom, 1977]. Initially, the cells deform to a bell type geometry. After some time the cells can collapse irreversibly onto the substrate and are then removed only with difficulty. The initial stage is predominantly one of elastic deformation. The subsequent collapse of the cell implies a dissipation of elastic energy and leads one to expect some metabolic or structural change in the cell. Thus this involves a mechanism whereby surfaces can stimulate energy dissipative processes within living organisms.

For homogeneous solids, quantitative theories exist which describe deformation extent in terms of the elastic constants of the solids and their interfacial contact energies. In general, the more strongly attracting the forces are and the 'softer' the materials the more pronounced is the deformation. Bacteria appear not to deform as readily as mammalian cells but often cause the cell-walls of tissues to deform. When in contact with, for example, metal surfaces, which give rise to very strong attachments, it is likely that bacteria will deform on contact.

The prevention of excessive contact deformation is probably an essential requirement for the maintenance of metabolic functions within attached living organisms. In this context, electrostatic forces or solvent mediated forces play a particularly important role, counteracting attractive dispersion forces and thereby reducing deformation pressures. Thus the function of surface electrical charges may not only be to mediate the long range aspects of interactions. Their more important biological function may be to ensure a minimal disruption of the internal structure of attached living organisms.

CONCLUSIONS

In this review we have outlined the main 'physical' forces to be considered in the adhesion of particles to substrates. In our view it is unrealistic to expect the current understanding of these to permit a comprehensive description of the many aspects of bacteria-substrate adhesion. It should be noted, however, that the existing theories of physical forces can give reasonable descriptions of the interactions between contacting, inert media of known composition. This is certainly true for long range forces, and significant advances have recently been made in theoretical descriptions of forces at short range.

It is to be expected that, as the complex architecture of biological surfaces is increasingly characterised, the physical treatments will become more comprehensively exploitable. A quantitative assessment might then be possible of the mediating roles of 'chemical bonds' (where expected) and of 'physical' forces. Though the former are usually referred to as 'specific', physical forces, to the extent that they reflect subleties of cell-wall architecture, can cause considerable adhesive specificity. Caution has to be exercised therefore in exclusively attributing specific effects to the consequences of specific chemical bonds, although at the same time, the possibility of chemical linkages has to be recognised.

We are grateful to Dr. A.E. Lee, Dr. J. Mingins and Dr. P. Richmond for helpful discussions.

REFERENCES

Bell, G.M. and Peterson, G.C. (1972). Calculation of the electric double-layer force between unlike spheres. *Journal of Colloid and Interface Science* 41, 542-566.

Chan, D., Mitchell, D.J. and White, L. (1975). Phase transitions in adsorbed polymer systems. *Discussions of the Faraday Society* 59, 181-188.

Clark, A.T., Lal, M., Turpin, M.A. and Richardson, K.A. (1975).

Configurational state of adsorbed chain molecules. *Discussions of the Faraday Society* 59, 189-195.
Curtis, A.S.G. (1973). Cell adhesion. *Progress in Biophysics* 27, 315-386.
Dahneke, B. (1972). The influence of flattening on the adhesion of particles. *Journal of Colloid and Interface Science* 40, 1-13.
Dahneke, B. (1975). Kinetic theory of the escape of particles from surfaces. *Journal of Colloid and Interface Science* 50, 89-107.
Derjaguin, B.V. (1934). Untersuchungen uber die reibung und adhasion, IV. Theorie des anhaftens kleiner teilchen. *Kolloid Zeitschrift* 69, 155-164.
Derjaguin, B.V. and Landau, L. (1941). Theory of the stability of strongly charged lyophobic sols and of the adhesion of strongly charged particles in solutions of electrolytes. *Acta Physicochimica URSS* 14, 633-662.
Derjaguin, B.V., Muller, V.M. and Toporov, Yu. P. (1975a). Effect of contact deformations on the adhesion of particles. *Journal of Colloid and Interface Science* 53, 314-326.
Derjaguin, B.V., Muller, V.M. and Toporov, Yu. P. (1975b). Influence of contact deformations on particle adhesion. 2. Macroscopic calculation of adhesive force with allowance for contact deformations in a spherical elastic particle. *Colloid Journal of the USSR* 37, 962-969.
DiMarzio, E.A. and McCrackin, F.L. (1965). One dimensional model of polymer adsorption. *Journal of Chemical Physics* 43, 539-547.
Dzyaloshinskii, I.E., Lifshitz, E.M. and Pitaevskii, L.P. (1961). The general theory of van der Waals forces. *Advances in Physics* 10, 165-209.
Evans, R. and Napper, D.H. (1977). Perturbation method for incorporating the concentration dependence of the Flory-Huggins parameter into the theory of steric stabilisation. *Journal of the Chemical Society Faraday Transactions I* 73, 1377-1385.
Gingell, D. and Fornes, J.A. (1975). Demonstration of intermolecular forces in cell adhesion using a new electrochemical technique. *Nature, London* 256, 210-211.
Greig, R.G. and Jones, M.N. (1976). The possible role of steric forces in cellular cohesion. *Journal of Theoretical Biology* 63, 405-419.
Hamaker, H.C. (1937). The London - van der Waals attraction between spherical particles. *Physica* 4, 1058-1072.
Hesselink, F. Th. (1977). On the theory of polyelectrolyte adsorption. *Journal of Colloid and Interface Science* 60, 448-466.
Hesselink, F. Th., Vrij, A. and Overbeek, J.Th.G. (1971).

On the theory of the stabilisation of dispersions by adsorbed macromolecules. II. Interactions between two flat plates. *Journal of Physical Chemistry* 75, 2094-2103.

Hill, T.L. (1960). *Introduction to Statistical Mechanics*, Chapter 15. London : Addison-Wesley.

Hogg, R., Healey, T.W. and Fuernstenau, D.W. (1966). Mutual coagulation of colloidal dispersions. *Transactions of the Faraday Society* 62, 1638-1651.

Johnson, K.L., Kendall, K. and Roberts, A.D. (1971). Surface energy and the contact of elastic solids. *Proceedings of the Royal Society* A324, 301-313.

Krupp, H. (1967). Particle adhesion theory and experiment. *Advances in Colloid and Interface Science* 1, 111-239.

LeNeveu, D.M., Rand, R.P. and Parsegian, V.A. (1976). Measurement of forces between lecithin bilayers. *Nature, London* 259, 601-603.

LeNeveu, D.M., Rand, R.P., Parsegian, V.A. and Gingell, D. (1977). Measurement and modification of forces between lecithin bilayers. *Biophysical Journal* 18, 209-230.

Lifshitz, E.M. (1955). The theory of molecular attraction forces between solid bodies. *Journal of Experimental and Theoretical Physics of the USSR, Zhurnal Eksperimental'noi i Teoreticheskoi Fiziki* 29, 94-110.

Lifshitz, E.M. (1956). Theory of molecular attraction forces between solids. *English Translation of Zhurnal Eksperimental'noi i teoreticheskoi Fiziki* 2, 73-83.

Marcelja, S. and Radic, N. (1976). Repulsion of interfaces due to boundary water. *Chemical Physics Letters* 42, 129-130.

Marshall, K.C. (1976). *Interfaces in Microbial Ecology.* Cambridge, Mass. : Harvard University Press.

Ninham, B.W. and Parsegian, V.A. (1971). Electrostatic potential between surfaces bearing ionisable groups in ionic equilibrium with physiological saline solutions. *Journal of Theoretical Biology* 31, 405-428.

Ohshima, H. (1977). A model for the electrostatic interaction of cells. *Journal of Theoretical Biology* 65, 523-530.

Okada, T.S., Takeichi, T., Yasuda, K. and Ueda, M.J. (1974). The role of divalent cations in cell adhesion. *Advances in Biophysics* 6, 157-181.

Parsegian, V.A. (1973). Long-range physical forces in the biological milieu. *Annual Review of Biophysics and Bioengineering* 2, 221-255.

Parsegian, V.A. and Gingell, D. (1973). A physical force model of biological membrane interaction. In *Recent Advances in Adhesion*, pp. 153-190. Proceedings of the American Chemical Society Symposium held in Washington DC, September 1971. Edited by L.H.Lee. London : Gordon and Breach.

Parsegian, V.A. and Ninham, B.W. (1970). Temperature dependant van der Waals forces. *Biophysical Journal* 10, 664-674.
Rees, D.A., Lloyd, C.W. and Thom. D. (1977). Control of grip and stick in cell adhesion through lateral relationships of membrane glyoproteins. *Nature, London* 267, 124-128.
Richmond, P. (1975). The theory and calculation of van der Waals forces. In *Colloid Science,* vol. II, pp. 130-172. Edited by D.H. Everett. London : The Chemical Society.
Richmond, P. (1977). Some fundamental concepts in flotation. *Chemistry and Industry* 792-796.
Rubin, R.J. (1965a). Random walk model of chain polymer adsorption at a surface. *Journal of Chemical Physics* 43, 2392-2407.
Rubin, R.J. (1965b). A random walk model of chain polymer adsorption at a surface II. Effect of correlation between neighbouring steps. *Journal of Research of the National Bureau of Standards B69,* 301-312.
Rubin, R.J. (1966). A random walk model of chain polymer adsorption at a surface. III. Mean square end-to-end distance. *Journal of Research of the National Bureau of Standards B70,* 237-247.
Smitham, J.B., Evans, R. and Napper, D.H. (1975). Analytical theories of the steric stabilisation of colloidal dispersions. *Journal of the Chemical Society Faraday Transactions I* 71, 285-297.
Spielman, L.A. and Friedlander, S.K. (1974). Role of the electric double layer in particle deposition by convective diffusion. *Journal of Colloid and Interface Science* 46, 22-31.
Van Oss, C.J., Good, R.J., Neumann, A.W., Wiesser, J.D. and Rosenberg, P. (1977). Comparison between attachment and detachment approaches to the quantitative study of cell adhesion to low energy solids. *Journal of Colloid and Interface Science* 59, 505-515.
Verwey, E.J.W. and Overbeek, J.Th.G. (1948). *Theory of the Stability of Lyophobic Colloids.* Amsterdam : Elsevier.
Weiss, L. (1974). Studies on cellular adhesion in tissue culture. XIV. Positively charged surface groups and the rate of cell adhesion. *Experimental Cell Research* 83, 311-318.
Weiss, L. and Harlos, J.P. (1972). Short-term interactions between cell surfaces. *Progress in Surface Science* 1, 355-405.
Wiese, G.R. and Healy, T.W. (1970). Effect of particle size on colloid stability. *Transactions of the Faraday Society* 66, 490-499.

ADHESION OF MICROORGANISMS TO SURFACES: SOME GENERAL CONSIDERATIONS OF THE ROLE OF THE ENVELOPE

H. J. ROGERS

*National Institute for Medical Research,
Mill Hill, London, NW7 1AA.*

INTRODUCTION

The importance of the ability of microorganisms to adhere to surfaces can hardly be overestimated in understanding their ecological behaviour. The proper functioning of agriculturally productive soil, the erosion of materials immersed in salt and fresh water and the virulence of some bacteria as pathogens for animals or plants, to cite only three examples, all depend in part upon the ability of microorganisms to adhere to the relevant surfaces. A wealth of detailed observation has accumulated to show, not only this, but that very considerable specificity is involved in the phenomena [see, for example, Smith, 1977; Gibbons, 1977], both as to the types of the organisms that will adhere to any given surface and the range of effective surfaces for any given microorganism.

Apart from these more commonly accepted roles for adhesive behaviour many other important events are also probably aspects of the same basic phenomenon. Among these may be mentioned the agglutination of microorganisms, the joining of bacteria during conjugation, the absorption of bacteriophage, the maintenance of ordered protein cell coats and perhaps the initial steps of DNA interaction with competent bacteria during transformation.

Despite the recognised biological importance of adhesion our knowledge of the underlying mechanisms is still woefully deficient. Attempts to analyse the phenomenon are not helped by the essentially qualitative nature of the term and the difficulty of designing methods by which to measure the degree to which even inorganic particles stick to each other [see Israelachvili and Tabor, 1973]. Nevertheless, valuable work has been done in biology particularly with mammalian cells to try to do this and to analyse the types of interactions involved [see, for example, Weiss, 1968; Weiss and Harlos, 1972; Baier, Shafrin and Zisman, 1968].

In this article I shall attempt to develop some general ideas about the adherence of microorganisms to surfaces and to each other. This will be done in relation to our knowledge about the structure and properties of the polymers known to make up the external walls of microorganisms as well as to those that are secreted through the wall into the external environment to form capsules or 'slimes'. No attempt will be made to catalogue the many known examples of bacteria sticking fast to surfaces or to each other.

THE NATURE OF FORCES BETWEEN PARTICLES

Any suspended particles whether of an inorganic or biological nature are subject to van der Waals forces of attraction. If in addition the particles have superficially located ionogenic groups and the suspending fluid contains polar, ionising substances, as is usual in biological situations, they will be subject to the forces arising from the interaction of the charged electric double layers around them as well as from other interactive electrostatic forces. The double layers arise because the fixed ionogenic charges on the particle, (for microorganisms these are usually COO^-, PO_4^{3-} and NH_3^+ groups) are neutralised by an atmosphere of oppositely charged ions drawn from the suspending fluid. Other types of electrostatic interaction are also operative and have been discussed and allowed for in the various calculations that have been made [Pethica, 1961; Weiss and Harlos, 1977].

The analysis of these long range forces between cells and between cells and surfaces is traditionally derived from the so-called DLVO (Derjaguin-Landau-Verwey-Overbeek) theory of colloidal stability originally developed independently by Derjaguin and Landau [1941] and Verwey and Overbeek [1948]. However, as will be seen, it seems unlikely that these sorts of forces of interaction although very important, can usually be sufficiently strong alone, to account for all adhesive phenomena. Hydrogen and ionic bond formation should also be considered. Cooperative interaction between numbers of such bonds would undoubtedly provide the necessary force as they do in providing much of the strength of natural products such as cotton fibres, chitin, gels of various sorts and many adhesives. That numbers of such bonds must cooperate to give sufficient strength could provide an explanation for the specificity of adhesion. The necessity for the correct positioning of donor and acceptor groups on the microbial cell and on the surface to which it will stick is likely to demand rather specific chemical structures.

LONG-RANGE VAN DER WAALS ATTRACTIVE FORCES AND IONIC REPULSIVE FORCES

As the name betrays, the original thinking about the formulation of van der Waals forces was concerned with an attempt to explain why real gases did not obey the simple rule learnt by school children that PV = RT. The analysis of the interaction between cells and surfaces in these terms has been modelled on those between inorganic particles in colloidal suspensions. Cells are, of course, vastly more complicated in shape and in non-uniformity of their surface molecular structure which makes the exact evaluation of the various functions in the developed equations extremely difficult - indeed at present impossible. Nevertheless, an estimate of the order and variation of force with decreasing distance between cells and surfaces can be obtained. The potential energy (V_T) between bodies can be formulated as the sum of the dispersive attractions and electrical repulsions as:

$$V_T = V_A + V_R$$

where V_A is the attractive dispersion force and V_R is the repulsive force.

This general equation has been further analysed by a number of workers interested in biological adhesion over the last fifteen years [Pethica, 1961; Weiss, 1968; Brook, Millar, Seaman and Vassar, 1967; Greig and Jones, 1976; Weiss and Harlos, 1972, 1977]. Because of the relatively large 'radius' of cells the analysis can effectively be in terms of two flat plates, two large spheres or a sphere and a plate. The choice leads of course to different equations [Israelachvili and Tabor, 1973], but the qualitative form of the curves obtained for interaction at decreasing distances is not necessarily different. The dispersive van der Walls energy of interaction (V_A) between two spheres is given by the equation [Israelachvili and Tabor, 1973; Weiss and Harlos, 1977]:

$$V_A = \frac{A}{6} \cdot \frac{r_1 r_2}{r_1 + r_2} \cdot \frac{1}{H}$$

where r_1 and r_2 are the radii of the spheres and H is their distance of closest approach. A is the Hamaker constant which is derived from the original London constant (C) in the equation for the dispersion energy (U) between two atoms in a gas ($U = C.d^{-6}$ where d is the interatomic distance). The Hamaker constant is obtained from C by summing the numbers of polarisable molecules per unit volume of the spheres interacting.

A recent attempt to evaluate this constant [Weiss, Nir, Harlos and Subjeck, 1975] for interacting mammalian cells has given a value of 5×10^{-14} ergs (5×10^{-21}J).

The repulsive energy between the spheres is more complicated when account is taken both of the overlapping ionic atmospheres and of the electrostatic interactions between the charged groups on the surfaces. If the spheres are regarded as maintaining a constant potential as they approach each other, one relationship, deduced by Hogg, Healy and Fuerstenau [1966] for colloidal suspensions is:

$$V_R = \frac{\varepsilon}{4} \frac{r_1 r_2}{r_1 + r_2} [(\psi_1 + \psi_2)^2 \ln(1 + e^{KH}) + (\psi_1 - \psi_2)^2 \ln(1 - e^{KH})]$$

where ε is the dielectric constant for the surrounding fluid which may differ significantly from that for water when biological fluids or growth media are involved. ψ_1 and ψ_2 are the surface potentials of the cells, usually taken as equal to the zeta potentials which are deducible from the electrophoretic mobilities of the cells. This may be an oversimplistic assumption in the case of bacteria which have a 'fuzzy' surface. $1/K$ is the Debye-Huckel parameter indicating the thickness of the ionic atmosphere around the bodies. A point to be noted, which will also be mentioned later, is that both V_A and V_R are closely related to the 'radii' of the cells.

When the computed interaction energies either between cells or between cells and plates are plotted against the distance separating them, curves of the general type shown in Fig.1 are obtained. Two characteristics of these curves are relevant. At short distances, of about 1 nm, between surfaces there is a very high potential energy barrier. This energy peak has a value of the order of 100 kT (4×10^{-12} ergs) for cells the size of bacteria. If there is to be direct interaction, this barrier must be crossed and energy either of locomotion or bombardment by other molecules must be supplied: calculations of the energy developed by a pseudomonad sufficient to drive it forward at 33 μm/sec has been found to be quite insufficient to do this [Marshall, Stout and Mitchell, 1971b]. Likewise molecular bombardment, imparting the familiar Brownian movement to bacteria, has been calculated to be about 1.5 kT and is also clearly insufficient [Brooks *et al.*, 1967]. Thus the bodies of undeformable cells are probably unable to pass the energy barrier to make use of the strong attractive forces that lie at even shorter distances.

A secondary small energy minimum, however, may exist at a much further distance between bodies of 5-10 nm (as in Fig.1). The forces here are rather weak and are not likely to account for the very strong interaction usually referred to by biolo-

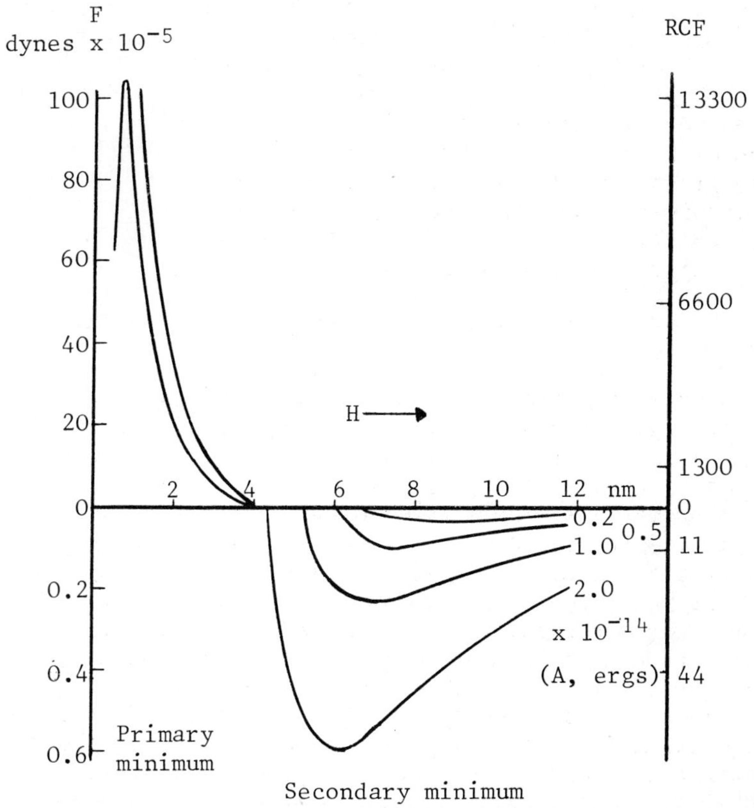

Fig. 1. The forces between a glassplate and a mammalian cell at different distances (H) from each other. The various curves are for different assumed values of the Hamaker constant (A). The figures under RCF on the right-hand axis represent the determined relative centrifugal forces required to overcome the forces of interaction. Taken from Weiss (1968).

gists as adhesion. They may however have the very important function of halting the cell long enough for other events to happen (that is they give rise to 'reversible' adhesions), so that hydrogen and ionic bonds can be formed by cell extensions or deformations. This cannot happen unless means of bridging the gap of 5-10 nm between the surfaces can be found, since the length of such bonds is of the order of only 0.2-0.3 nm. Indeed in the work on a pseudomonad already mentioned [Marshall *et al.*, 1971a, b], the microorganisms were held for a time at a distance from the surface whilst, when motile, they rotated by flagellar action. They could be readily removed from such a situation by washing the surface with saline and this phenomenon was referred to as reversible adhesion.

As pointed out, the energy of interaction between spheres or between a sphere and a plate is related to the radius. As the radius of the sphere is reduced, the potential energy barrier becomes progressively smaller. Fig. 2 (taken from Weiss and Harlos, 1977) shows the potential energy of interaction between spheres of different diameters and for a body with a radius of 0.05-0.1 μm the barrier is of the order of only 10^{-13} ergs. This value is sufficiently small enough to be overcome either by forces of locomotion developed by microorganisms or molecular bombardment. Thus if the cell has, or can extrude, probes of narrow diameter, these can easily come within close enough range of a surface for hydrogen or ionic bonds to be formed. When considering bacteria it is necessary to remember the presence of fimbriae, pili and flagellae any of which would theoretically provide ideal probes. Likewise deformation of mammalian cells may equally be effective.

A second way in which the potential energy barrier between cells and surfaces can be circumvented is for the microorganism to form extracellular material (say polysaccharide) which as polymeric molecules will again be able to approach close to the surface. If the surface happens to be that of a cell in a mammalian tissue, this too is likely to have polysaccharide in the form of a glycoprotein coat, outside its cytoplasmic membrane. Such an extracellular coat contains polymers able to form bonds either with the surface of the microbe or with extracellular polymers produced by it.

When coats of 'soluble' polymers surround microorganisms, considerations of the wettability of the substrate by the polymers must also arise [Baier, 1970, Baier *et al.*, 1968]. If two smooth surfaces are brought together with a layer between them of liquid which wets them (that is one with a low contact angle to the surfaces) (see Fig. 3), they will strongly resist attempts to pull them apart, although they can easily slide across each other (that is they have little resistance to shear). If the fluid layer is viscous or is a gel, there will also be resistance to shear in proportion to the viscosity of the sandwich of fluid.

One further type of mechanism of interaction between microbes and surfaces may be illustrated by the immobilisation of bacteria, instead of enzymes on insoluble supports. It has been reported [Kennedy, Barker and Humphreys, 1976] that *Escherichia coli* and *Saccharomyces cerevisiae* can be fixed to the surface of zirconium hydroxides. The authors suggest that the

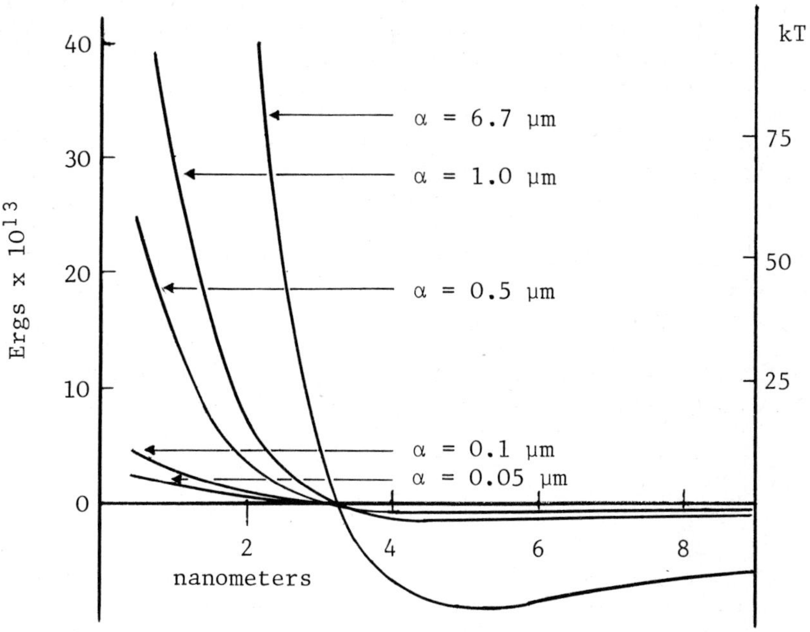

Fig. 2. The computed energies of interaction at different distances between the flattened region of a cell and probes from another cell of different radius of curvature. The right-hand axis is in units of potential energy, where k *is the Boltzmann constant and* T *the absolute temperature; on the left the force in ergs is computed (from Weiss and Harlos, 1977).*

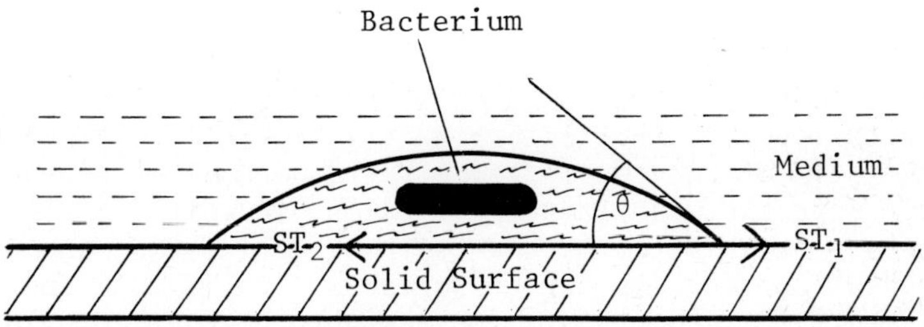

Fig. 3. A model of a bacterium sticking to a solid surface by the glue of a capsular polysaccharide. θ *is the contact angle and* ST_1 *and* ST_2 *are the surface forces between the concentrated capsular polysaccharide solution and the medium* (ST_1)*, and the capsule and the solid surface* (ST_2).

hydroxide as they prepared it, was likely to have the following structure.

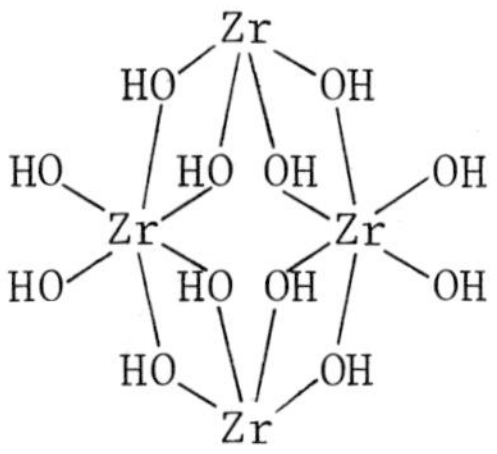

Enzymes or the surface polymers in the organisms having ligands such as -OH, $-COO^-$ or NH_3^+ groups would form partial covalent bonds with some of the -OH groups in the hydroxide. If this explanation of the retention of the microorganisms is correct, it must be assumed that close enough contact can be established owing to the electrostatic charge characteristics of the hydroxide and because the latter is essentially 'molecular' in dimensions, and not polymerised. As the authors say almost any assumption about the exact nature of the surface of the microbes is likely to provide plenty of appropriate ligand groups.

THE NATURE OF THE BACTERIAL SURFACE IN RELATION TO THE TOTAL POTENTIAL ENERGY EQUATION

From what has already been said, it would seem unlikely that our knowledge of the bulk chemistry of 'clean' wall preparations, or with Gram-negative organisms, of components of the outer envelope, will necessarily help very much in our understanding of the phenomenon of adhesion. What is needed is a knowledge of the molecular and ionogenic organisation of the outermost surface of the cell and of its appendages. Three methods are relevant. These are firstly immunological methods and secondly, the location and nature of bacteriophage acceptors. In neither case would evidence suggest that the reagents (that is serum antibodies or phages) are able to penetrate into the bulk of the 'wall'. The third method is to measure the electrophoretic mobility of the cells. Direct measurements such as by the latter method can provide information about the averaged ionogenic nature of surfaces - their zeta and hence their surface potentials. They do not, however, provide information about the topological distribution of the charges. By chemical manipulation of the wall polymers, followed by microelectrophoresis it has sometimes been possible to learn about the molecular structure of the compounds bearing the ionogenic groups.

Immunological methods and bacteriophages for exploring the surface

The use of either method brings to light a major difficulty in defining the cell surface. When microorganisms are examined either with the light microscope or by the electron microscope the impression given is of a fairly well defined external boundary. The preparation and examination of walls, or in the case of Gram-negative species, fractions constituting the envelope, has tended to confirm the idea of the presence of a clearly defined entity, the 'wall', surrounding the bacteria. When the surface is probed instead with large molecules such as antibodies or particles, for example, bacteriophages, which can combine specifically with entities at a distance from the usually recognised boundary of the wall without being able to penetrate into them, the surfaces are seen to be organised in depth. Materials of different structure may extend and interpenetrate each other, well beyond what is usually regarded as the edge of the wall. A few examples will illustrate these points.

Antisera made by injecting animals with whole Gram-negative bacteria may contain antibodies to lipopolysaccharides (LPS), microcapsular antigens such as K and Vi, pili, fimbriae and flagellae. Those reacting with the lipopolysaccharides of *Salmonellae* for example have exquisite specificities for the structure of the carbohydrate side chains and cores of the molecules and this gives rise to the large number of serotypes of these organisms. The LPS are exclusively associated with the outer membrane of the envelope when this is isolated and might be thought of as a characteristic of this structure which, as seen in transverse sections of the bacteria, has a fairly clearly defined boundary. However, if ferritin-conjugated antibodies are used to treat the cells before sectioning it can be seen that in some microorganisms, such as *Escherichia coli* and *Salmonella typhimurium*, the LPS extends out beyond the outer membrane by as much as 150 nm [Shands, 1966]. Such a consideration is clearly important when considering the interaction of bacteria with surfaces.

Other antigens of a polysaccharide nature, such as the common reacting antigen [Makela and Mayer, 1976] are also associated in some way with the outer membrane so that they react with antibodies which, unlike those of the LPS do not agglutinate the organisms. Antisera to whole organisms contain other antibodies, such as those to microcapsular polysaccharides with which whole organisms also react and it is necessary to explain how all these antigens can be on the surface simultaneously assuming the surface to have a clearly defined 'edge'. Some patchiness would seem to be a possibility but reaction of the organisms with antibodies conjugated either with fluorescent substances or ferritin show that this is unlikely to be so. A more likely alternative is that the surface layers are not clearly defined but represent a loosely organised mixture of polymers which might be looked at as a highly concentrated gel which is penetrable to varying extents by antibody molecules.

In Gram-positive bacteria the thick wall is made up of cross-linked peptidoglycan covalently linked with highly negatively charged polymers such as teichoic and teichuronic acids. Evidence suggests [Scherrer and Gerhadt, 1971], that the porosity of the walls is such that globular proteins, like antibodies, cannot penetrate into them. Again the use of immunological methods suggests that the surface layers, such as capsules, are likely to be built in depth through which the protein molecules present in antisera can usually move. The walls themselves containing as they do more than one polymer could have all, none or some exposed at the surface.

Antisera to a strain of *Bacillus licheniformis* contained reactive antibodies to both teichoic acid and peptidoglycan although those to the former predominated. However, antibodies made against isolated peptidoglycan, free of teichoic acid, agglutinated or reacted with whole organisms [Hughes, Thurman and Salaman, 1971; Hughes and Stokes, 1971]. Thus both wall polymers are presumably exposed at the 'surface'.

In *Bacillus subtilis*, a closely related organism, evidence has been obtained [Burger, 1966] suggesting that the teichoic acid is mostly buried beneath other immunogenic material and can only be made available by partial enzymic hydrolysis of the peptidoglycan.

Some lactobacilli and streptococci illustrate a different point of considerable interest. In these organisms there is good evidence that material normally reasonably thought to be associated with the cytoplasmic membrane which lies deeply buried under a wall, 30-40 nm thick, can penetrate through to the exterior surface. A well-known and classical example of this is the agglutination of Group D streptococci by antisera with specificity directed towards the membrane associated lipoteichoic acid [Shattock and Smith, 1963; Wicken, Elliot and Baddiley, 1963].

In Group A streptococci, a similar situation is suggested in work directly relevant to the subject of adhesion. It has been commonly assumed that because treatment of the bacteria with trypsin removes simultaneously their coronae of coarse fimbriae-like protusions, their M reactive protein and their ability to adhere to tissue cells, that the latter two phenomena are causally connected [Swanson, Hsu and Gotschlich, 1969; Smith, 1977]. This assumption has been called into question by the finding that treatment of the bacteria with very dilute pepsin at pH 5.8 removes their M protein but not their coronae, neither does it stop them from adhering to epithelial cells [Beachey and Ofek, 1976]. Adhesion was on the other hand, inhibited by antibodies to the purified lipoteichoic acid isolated from them [Ofek, Beachey, Jefferson and Campbell, 1975]. Treatment of the epithelial cells either with the lipoteichoic acid or with the glycolipid isolated from it also stopped adhesion. This provides further evidence that the membrane associated teichoic acid penetrates through the wall into the exterior medium and can be important for adhesion by the cells.

Lactobacillus fermentii is readily agglutinated by antisera directed against lipoteichoic acid, whereas the related *Lactobacillus casei* is not [van Driel, Wicken, Dickson and Knox, 1973]. The use of ferritin-labelled antibodies, however, showed that this latter organism also reacts with the antibody.

The reverse situation can also sometimes apply. Large important amounts of material sometimes even visible under the light microscope at the surface of microorganisms can be 'invisible' to antibodies. For example, some Group A and C streptococci are known to be surrounded by large capsules of hyaluronic acid which have a profound effect upon the surface potential of the cells (Fig.4), [Hill, James and Maxted, 1963] and which are highly likely to modify their interaction with surfaces. Hyaluronic acid is non-antigenic like teichuronic acid in the walls of bacilli and does not therefore produce antibodies. Nor do the capsules prevent antibodies directed towards the cell wall antigens reaching and reacting with them. Likewise, the presence of the M antigen does not stop antibodies reaching the C polysaccharide that is covalently attached to the peptidoglycan; neither does the teichuronic acid in the walls of bacilli prevent antibodies reacting with either teichoic acid or peptidoglycan.

The distribution of receptor sites for bacteriophages in both Gram-positive and Gram-negative bacteria has been very thoroughly discussed by Lindberg [1973]. In both *Escherichia coli* and *Salmonellae* phages can be found with specificities directed with great precision to the different side chains of the LPS molecules as well as to their core structures. Many phages such as T_1, T_5, T_6 and Ø80 also react with proteins in the outer membrane of the envelopes of the organisms. Evidence, however, would suggest that the approach of some of the core specific phages to their receptors is hindered even by the presence of the O side chains of the LPS, despite the fact that other phages can travel through the coat of LPS to reach their protein receptors in the outer membrane of the envelope.

The capsular polysaccharides through which LPS may or may not penetrate, such as the Vi antigen, can also act as phage receptors from which the particles sometimes liberate them-

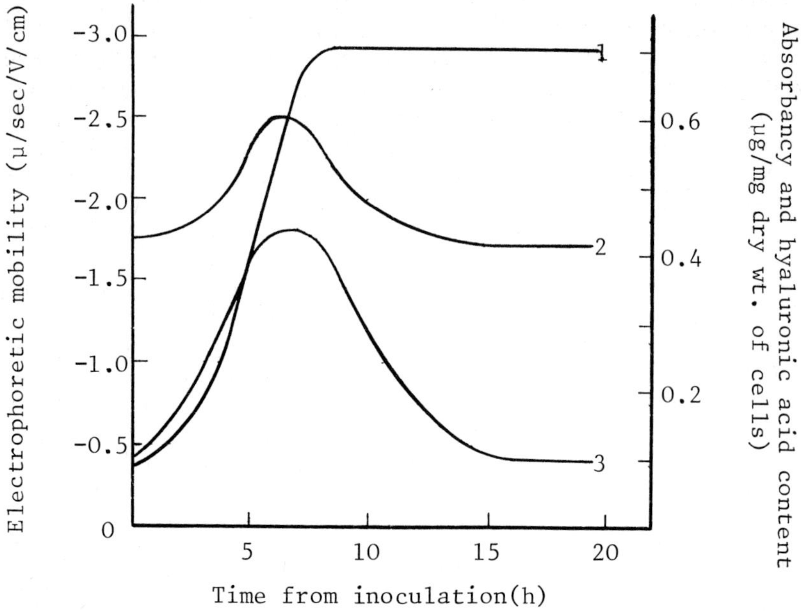

Fig. 4. The variation of the electrophoretic mobility and hyaluronic acid content with age during the growth of Type 12G Streptococcus pyogenes. 1 - *growth curve;* 2 - *electrophoretic mobility;* 3 - *hyaluronic acid content (from Hill et al., 1963).*

selves by deacetylating a high proportion of the *N*-acetyl-D galactosaminuronic acid of which the antigen is made. It would seem possible that in such examples absorption to the capsular polysaccharide represents a temporary halt on the way to receptors elsewhere in the outer layers but that the polysaccharide presents too physical a barrier. In other instances, evidence has been produced to suggest that excess of capsular material may obstruct some phages with specificities directed to the LPS, but not others from reaching their receptors.

Apart from the specificity for capsular materials and components of the outer layers of the envelopes of Gram-negative bacteria, other phages can be specific for extensions of the organisms such as pili and flagellae, differentiation even being expressed between the sex pilus and other pili [Ottow, 1973]. Thus by choice of the right bacteriophage the outer regions could be defined in depth. The adsorption of bacteriophages by their receptors among Gram-negative bacteria has been used here simply to illustrate the organisation of the superficial layers of host bacteria. The subject could have been used to illustrate the complexities of the whole problem of adhesion in biological systems, had space permitted. Work on this subject is highly relevant to the present topic.

Among Gram-positive species evidence from *Staphylococci*, *Streptococci*, *Micrococci* and *Bacilli* would all show that phage receptors can frequently involve wall teichoic acids or polysaccharides but these must be covalently linked to peptidoglycans. One particularly relevant study by Coyette and Ghuysen [1968] showed that some staphylococcal phages can fix to a soluble complex, consisting of teichoic acid linked to a glycan strand from the peptidoglycan. This suggests that both the teichoic acid and in particular its glycosyl substituents, or β-*N*-acetylglucosaminyl groups, together with considerable lengths of the glycan strands of the supportive peptidoglycan, are exposed at the surface of the walls. They then act together as phage acceptors.

Again, as in Gram-negative species, the presence of capsules outside the 'wall' may either hinder the approach of phages to deeper lying receptors, act as receptors themselves or have no apparent effect. An example in which phages may in future be used as probes for the most external surface of microbes is that of an extracellular layer in the form of an ordered protein coat of *Bacillus sphaericus* which acts as a location of phage receptors [Howard and Tipper, 1973]. In mutants with altered resistance to phages, the protein in the layer was modified chemically, but still inactivated other phages, suggesting multiple receptors in the protein coat which itself is held to the wall either by secondary valency or London forces similar to those involved in adhesion.

One advantage of both the immunological methods and the use of bacteriophage receptors to probe the surface, is that they can give some indication not only of the organisation of the surface in depth but of the topological distribution of the polymers involved. Fluorescent-labelled antibodies adsorbed on to bacteria and examined under the ultra-violet microscope, can only give a general impression. Ferritin-labelled antibodies do better, however, and within the limits imposed by the specificity of the antisera used and the size of the antibody molecules, may give much more precise indications of the distribution of antigens. Bacteriophages being much larger structures even though adsorbed by their tails, may involve larger and more heterogeneous areas of the surface such as would be suggested from the results with the soluble glycan-teichoic acid complex [Coyette and Ghuysen, 1968]. Nevertheless, examination under the electron-microscope after appropriate staining can show whether or not the distribution of receptors is random over the surface. In general where saturating doses of bacteriophage have been used, the surfaces of bacteria indeed appear to absorb phage in a random fashion (see, for example, Archibald and Coapes, 1976).

Electrophoresis of microorganisms

When suspended in appropriate buffers in a cell connected to electrodes, the rate of movement of microorganisms can be measured by observation under the microscope. This mobility is directly related to the ionogenic nature of the surface. Unlike the methods discussed above the average charge on the cell can be measured with no indications of the topology of the surface. The influence on the total surface potential of the organisation of the surface in depth, such as shown to be likely from the application of the other methods, is unclear. This could be one of the uncertainties in making calculations of the energies of interaction between cells and surfaces such as those that have been discussed in the earlier sections of this chapter.

All cells, including microorganisms, that have been examined have a net negative charge and surprisingly similar mobilities at pH 7.0, ranging around -1 to -3 x 10^8 $m^2s^{-1}V^{-1}$.

Results revealing more about the surface can be obtained by examination of mobilities after chemical or enzymological manipulation. For example, removal of hyaluronic acid capsules from the surface of a strain of *Streptococcus pyogenes* lowered the mobility from about -1.6 x 10^8 $m^2s^{-1}V^{-1}$ when the organisms had 0.24 μg/mg dry weight associated with them to -1.0 x 10^8 $m^2s^{-1}V^{-1}$ when they had no detectable polysaccharide. This fall in mobility was presumably directly due to the loss of the $-COO^-$ of the uronic acid in the hyaluronic acid [Hill *et al.*, 1963].

A more interesting result was obtained with *Staphylococcus aureus*. When the mobility of this bacterium was examined at different pH values, it was maximal between pH 3 and 4 indicating the presence at the surface, of anionic groups with low pK values. When treated for a short time with sodium metaperiodate at 0°C and pH 7.0, the picture was completely changed. At pH 3.2, the cells had no charge at all, slowly rising to a maximum at pH 7 to 8. Other work [Rogers, 1963; Garrett, 1965] had shown that the teichoic acid - a polyribitol phosphate in this organism - was oxidised and removed by such treatment with periodate. Thus it is reasonable to suggest that the large negative charge found at pH 3.4 was due to exposure of the ribitol teichoic acid at the surface of the staphylococci. Such a result would then confirm both immunological evidence and that on the distribution of bacteriophage receptors.

EVIDENCE AS TO THE MECHANISMS OF ADHESION

The basic hypothesis put forward at the outset is that microorganisms will not be likely to adhere directly either to each other or to surfaces over any considerable area of the cells by the operation of long range forces. This is due to the high energy barrier to be overcome resulting from their size and partly from the strongly ionogenic nature of the surfaces. If however, thin probes of soluble hydrophilic polymers such as polysaccharides are formed either by the microorganism or by the surface to which they are adhering, particularly if this should be that of a living cell, then strong bonds and forces may be operational.

When soluble polysaccharides are formed, adhesion may take place by surface forces or by the formation of specifically arranged hydrogen or ionic bonds. The 'surfaces' of many bacteria are organised in depth with a variety of polymers exposed many of which bear a preponderance of negative charges. Such 'layers' of the surface are penetrable to varying degrees by large molecules and it would appear that they are often loosely enough organised to be regarded as fundamentally highly concentrated solutions of polymers.

The first deduction from the theoretical considerations namely that direct contact between the bacterial wall and surface cannot be achieved, is verified by observation. When the morphology of bacteria adhering to eukaryotic cells has been examined (see, for example, Wagner and Barnett, 1974; Staley, Jones and Corley, 1969; Takeuchi, 1967; Cleveland and Grimstone, 1963-4), a gap of about 10 nm or more between the wall of the organism and the plasma membrane of the eukaryotic cell has always been recorded. In some instances but not others, this is occupied by what have been called striated capsules.

The role of probes

The probable role of fimbriae in the adhesion of Gram-negative bacteria to a surface, both living and dead, has been known for over twenty years [Duguid and Gillies, 1957, 1958; Duguid, Smith, Dempster and Edmunds, 1955]. Such structures form ideal probes from the physico-chemical points of view discussed earlier. They are often no more than 4-7 nm in diameter but up to 1000 nm in length; there are several hundred of them usually arranged peritrichously. They are therefore thin enough to 'evade' the potential energy minimum between the cell and the surface, and randomly arranged so that the stance the organism adopts to the surface is unimportant.

Two other observations in the early work are of relevance. Firstly fimbriated strains are very catholic in their tastes as to the surfaces to which they will adhere. Included in the early tests were red blood cells derived from a number of species, and the stromata from them, leucocytes, intestinal epithelial cells, microfungi, plant root hairs, glass and cellulose fibres. Secondly two classes of fimbriate adhesion were recognised. In one class adhesion to cells of all sorts was inhibited by the sugar mannose; adhesion to glass was not inhibited. Further work has produced other sub-divisions according to the behaviour of fimbriae [Ottow, 1975] and four types have been described (Table 1).

Table 1.

Classification of fimbriated bacteria

Sub-type	General adhesive ability	Inhibition by mannose[1]	Diameter (nm)	Attachment to animal cells
1	++	++	7.0	++
2	0	-	7	0
3	++	0	4.8	0
4	++	0	4.8	+

[1]Also inhibited by methyl-α-D-mannoside. This inhibition does not apply to adhesion to glass or cellulose.

The chemistry of pili or fimbriae from *Escherichia coli* has been examined and they have been shown to consist predominantly of a sub-unit of proteins, MW about 16,000, now called pilin and fimbrilin. Recently Ofek, Mirelman and Sharon [1977] have started to examine the chemical mechanism of the adhesion of *Escherichia coli* to epithelial cells which is inhibited by mannosides and have suggested the presence of a lectin-like substance on the surface of this organism. They succeeded in extracting a substance with buffered saline at pH 8.5 that inhibited attachment to epithelial cells. The extracts also agglutinated yeast to which it adsorbed. Treatment of the epithelial cells with concanavalin A or mild oxidation with periodate prevented subsequent adhesion of *Escherichia coli*. Whether the extracts from the bacteria contained active entities from the fimbriae or pili or even possibly contained fimbriae themselves is not clear.

The combined evidence suggests that there is a specific interaction between fimbriae and mammalian cell surfaces in those species in which adhesion is inhibited or reversed by mannose or α-methyl-D-mannoside. However, early work showed that such cells also adhered to glass but that this was not inhibited by the sugar. It is thus possible that both specific and non-specific adhesion of the bacteria may be possible with the help of these small diameter probes. If this general hypothesis is correct then in specific adhesions, hydrogen and ionic bonds would be found between the polymers on the fimbriae. In the latter the London forces expressed in the primary energy minimum may be sufficient.

It should be pointed out that so far no specific inhibitors of adhesion by fimbriated microorganisms of sub-types three and four (see Table 1) have been described. This may either mean that the adhesion is non-specific or that the right inhibitors have not been found. It may be relevant that these two groups have fimbriae that are even thinner than those in the first two sub-types. The presence of sub-type 2, which has fimbriae of the same diameter as those in sub-type 1 but to not adhere to cells, raises the possibility of a thin coating of specific adhesive substances on some but not other fimbriae which has so far either not been detected by analysis of isolated fimbrilin, or has been lost in the process of isolating the protein.

One possible role of fimbriae that should be born in mind is their interaction with surfaces which could form a second stage in a three stage process. The first stage would be weak reversible adhesion due to the forces in the secondary energy minimum (see page 32). The second would be firmer non-specific adhesion by the fimbriae and pili involving the formation of some hydrogen or ionic bonds. The third stage would be the formation of extra-cellular material such as polysaccharide either by the cell, or by the bacterium. The first two stages would happen in very rapid succession of the order of milli- or even micro-seconds, whilst the third would be expected to happen very much more slowly dependent on the biosynthetic process of the cells and bacterium.

The role of pili in promoting the adherence of gonococci to mammalian cells has been much discussed [Swanson, 1975, 1977; Smith, 1977] and may indicate the sort of complexity just mentioned. It would appear that pili are either necessary or certainly strongly promote adherence of the microorganisms to *in vitro* grown mammalian cells of epithelial origin, erythrocytes and human sperm. However, it is claimed that both piliated and non-piliated laboratory grown bacteria adhered equally well to epithelial cells in human fallopian tubes and the endocervix. Pili of gonococci may apparently even be associated with the failure of the bacteria to adhere to mouse peritoneal macrophages, whereas non-piliated ones do so.

These complications may of course, suggest that the whole thesis advanced here is wrong. However, much more evidence is required about the chemical nature and potential of the surfaces of both the organisms, together with their pili and of the cell surfaces to which they either do or do not adhere, before drastic revision of these ideas should be undertaken. Even more important is perhaps the design of techniques to quantitate adherence and distinguish adherence from entanglement and engulfment when adherence to living cells is studied.

Apart from well recognised probes such as fimbriae and pili, other sorts of organised appendages are present on some bacteria. For example, *Flexibacter polymorphus*, a marine gliding bacterium, has an array of cup-shaped hollow goblets apparently attached to the LPS. From each of these structures a single filament of from 1.5 to 2.5 nm in diameter and as long as 1 μm in length emerges [Ridgway, 1972]. Such fibres would be ideally suited to help in sticking the organisms to surfaces; indeed they have been seen appearing to do so in scanning electron microscope pictures of the organism adhering to the surface of a membrane filter [Ridgway and Levin, 1973].

Another fascinating example of specialised probes is provided by the members of the genus *Simonsiella*. These organisms consist of filaments built up from large disc-like lunate cells. The filaments adhere and glide on their ventral sides and each of the discs on this side appears to have protruding from it, an array of fibres or bristles each about 300 nm long. The total appearance of the organism is of a worm standing on its chaetae (Fig.5) [Pangborn, Kuhn and Woods, 1977].

The role of surface glues

That extra-cellular polysaccharide or at least polymer formation by bacteria is involved in their ability to adhere to surfaces, is a well recognised fact [Fletcher and Floodgate, 1973; Gibbons and Houte, 1975; Marshall *et al.*, 1971b; Gibbons, 1977]. These polymers can be seen by the scanning electron microscope as fine fibrils appearing to bind the organisms to the surface of, in this case, columns of epithelium of the colonic mucosa of mice [Savage, 1977]. The situation in which say a polysaccharide is formed by a microorganism and remains firmly attached as a capsule is fairly easy to comprehend. It may simply form a glue or it can be imagined that the polysaccharide molecules form hydrogen bonds with substances on the surface to which adhesion occurs. Alternatively if it is an acidic polysaccharide, it may form ionic bonds with divalent metal ions involving a second array of negative charges in the surface.

A beginning has been made [Morris *et al.*, 1977] on studying the specificities involved in polysaccharide recognition

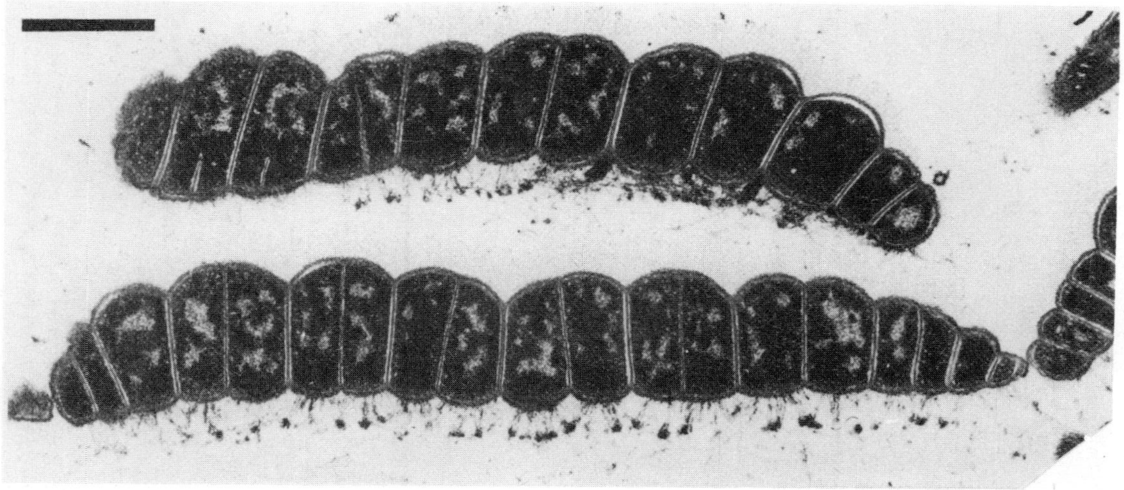

Fig. 5. Longitudinal section of *Simonsiella*, a cyanobacterium, showing the specialised structure between the ventral side of the organised collection of cells and the agar surface (taken from Pangborne et al., 1977).

with xanthan produced by *Xanthomonas*. Xanthan consists of β(1-4) linked glucose residues substituted by trisaccharide side-chains. The authors used a variety of spectroscopic techniques to study the conformation of the xanthan. They showed that the polysaccharide had a temperature induced transition from a random coil at higher temperatures to a helix in its native form, at lower temperatures. The strengths of gels formed in mixtures with galactomannans isolated from various sources was then studied. They were roughly proportional to the molar ratio of the mannose to galactose residues in the galactomannans.

This suggested to the authors that the helical form of the xanthan was able to form a strong association with those parts of the galactomannans where there were unbranched chains of mannose residues. The xanthan was also said to associate even more strongly with glucomannans and soluble cellulose derivatives. Association of this sort with insoluble polysaccharides on surfaces could clearly account for microbial adherence. Alternatively interaction between loosely associated polysaccharides could provide adequate shear resistant glues between microbe and surface.

Insoluble polysaccharides as mooring-lines

It would seem that insoluble polysaccharides produced by microorganisms are often involved in adhesion. Whether these function as a 'glue' does not seem to be likely. Rather interaction between the surface and the polysaccharide fibres would be predicted as more likely. A classical example is to be found in the adherence of *Streptococcus mutans* to glass surfaces. This organism synthesises a mixture of high molecular weight insoluble dextrans (glucans) and 'levans'. Living cells incubated with sucrose will adhere to glass sufficiently strongly that they can only be detached by rubbing the surface with a rubber covered rod. Heat-killed cells also adhere if they are incubated with a mixture of cell free enzymes that will synthesise the polysaccharide from sucrose which must also be added to the system. Such adherence is inhibited by antibody, specific for the cell-wall group polysaccharide or by treatment of the cells with enzymes such as pepsin, papain or dextranase [Mukasa and Slade, 1973; Slade, 1977; Kelstrup and Funder-Nielson, 1974]. Agglutination under similar circumstances is also inhibited by trypsin treatment [Spinnell and Gibbons, 1974].

The conclusion drawn from this work is that the biosynthetic enzymes are adsorbed to the wall of the organism by a soluble dextran (presumably formed by other biosynthetic enzymes either wall or membrane associated) together with a wall protein. In this situation the enzymes manufacture the mixture of insoluble polysaccharides that leads to adhesion of the bacteria to glass. Since adhesion is also inhibted by group specific antibody to the wall polysaccharide, it is assumed that this antibody obscures the sites for the total adsorption phenomenon.

One difficulty in interpreting all this work is that the bacteria were harvested from overnight cultures made in rich medium. It is, therefore, not wholly possible to escape the suspicion that the intra-cellular biosynthetic enzymes and other proteins might have leaked from some cells, dead or dying, and been adsorbed, on to the surface of other living cells. If this is not so then adherence of *Streptococcus mutans* involves a complicated double phenomenon. Adhesion of the insoluble polysaccharides both to the bacteria and to the glass surface. Even if leakage from dead cells is the explanation, it does not diminish the interest of the observations or their usefulness in understanding dental plaque formation where death of part of the population is also likely to occur.

The nature of sticking of the insoluble, neutral polysaccharide to glass also needs investigation. Although hydrogen bond formation may occur, it may also be that the fibres can approach closely enough to be held in the primary energy minimum by the van der Waals forces. On the other hand, it is possible that components of the medium, material leaking from the heated bacteria, or components of the crude enzyme preparation are first fixed to the glass and that the polysaccharide fibres can then interact with these layers (but see Fletcher, 1976).

A rather similar formation of fibrils has been described, during the 'irreversible' adhesion of a marine *Pseudomonas* sp. to glass or to plastic covered nickel grids [Marshall *et al.*, 1971b]. For this type of adhesion to occur both glucose and either Ca^{2+} or Mg^{2+} had to be present in the suspension of bacteria. When firm adhesion occurred, fibrils of material could be seen by examining the nickel grids under the electron-microscope. It is particularly interesting that where bacteria had been sheared from the grid 'marks' remained, looking like detached thin fibrils or impressions of fibrils.

Can lipopolysaccharides act as adhesives?

As has been pointed out, lipopolysaccharides extend a considerable distance from the end of the outer membrane of organisms such as *Escherichia coli* and *Salmonellae*. They could, therefore, have the role of reaching the surface to which the microbes adhere and 'bridge' the high energy barrier. An example [Wolpert and Albersheim, 1976] in which this may happen, is to be found in the interaction of *Rhizobia* with legumes and with lectins extracted from them. In this work the bacteria were extracted with phenol-water mixtures by the now traditional method for obtaining LPS. Hydrolysates of the extract were shown to contain components typical of LPS. When passed down affinity columns to which lectins extracted from a variety of legume seeds were fixed, about 30-40% of the LPS components became strongly adsorbed.

The most important point of this work was that the specificity interrelations between the *Rhizobia* and the plant species were preserved, (that is the LPS of the organism usually associated with the particular species of legume was fixed only to this particular lectin in the columns). However, since only 30% or so of the LPSs were so fixed, perhaps it is premature to assume that these substances are responsible for the adhesion involved on the living plant. Further work to test adhesion itself in relation to purified LPSs of known structure, is clearly important. In passing it may be noted that the interpretation of this work is exactly in the opposite sense to that of Ofek *et al.* [1977]. In the latter, the lectin was suggested as being on the bacteria and its complement was on the mammalian cell surface to which the bacteria adhered.

CONCLUSION

There would appear to be no doubt that the subject of the adherence of microorganisms to surfaces is very important and ripe for investigation. If the subject is to be advanced, it would seem that certain general aspects and approaches must be secured.

The most difficult aspect in assessing the very large body of natural history associated with the subject of adhesion, is the lack of quantitative data. Techniques ought to be designed in an attempt to measure in meaningful units the forces involved in the removal of microbes, both from inert surfaces and from cells. More fundamental work of the type described for the interaction of xanthan and galactomannans is also needed and should be extended in an attempt to distinguish between adhesion due to formation of glues, from that due to direct linkage by specific constellations of secondary bonds in lectin or enzyme-substrate like interactions. More work involving the design of new 'fast-flow' techniques is required, to study the time sequence of adhesion particularly with fimbriated organisms. Likewise, further morphological work is needed using as many preparative techniques for electron microscopy as are available, in order to try to define the exact situation of an organism when adhering to surfaces. Finally, the subject needs an injection of biochemical and genetic expertise, to try to examine further the nature of the polymers involved in adhesion. This applies equally to envelope associated and extracellular substances formed by microbes and to those on the surfaces of mammalian and plant cells.

The strongest impression gained from the literature is that it would be wrong to regard adhesion as universally due to a single mechanism. Different microorganisms use different means and any one organism may use several means at different times or when adhering to different surfaces.

REFERENCES

Archibald, A.R. and Coapes, H.E. (1976). Bacteriophage SP50 as a marker for cell wall growth in *Bacillus subtilis*. *Journal of Bacteriology* 125, 1195-1206.

Baier, R.E. (1970). Surface properties influencing biological adhesion. In *Adhesion in Biological Systems*, pp. 15-48. Edited by R.S. Manly. New York : Academic Press.

Baier, R.E., Shafrin, E.G. and Zisman, W.A. (1968). Adhesion: mechanisms that assist or impede it. *Science* 162, 1360-1368.

Beachey, E.H. and Ofek, I. (1976). Epithelial cell binding of Group A streptococci by lipoteichoic acid on fimbriae denuded of M protein. *Journal of Experimental Medicine* 143, 759-771.

Brooks, D.E., Millar, J.S., Seaman, G.V.F. and Vassar, P.S. (1967). Some physico-chemical factors relevant to cellular interaction. *Journal of Cellular Physiology* 69, 155-168.

Burger, M.M. (1966). Teichoic acids: antigenic determinants, chain separation and their location in the cell wall. *Proceedings of the National Academy of Sciences of the United States of America* 56, 910-917.

Cleveland, L.R. and Grimstone, A.V. (1963-1964). Fine structure of the flagellate *Mixotricha paradoxa* and its associated organisms. *Proceedings of the Royal Society, London*, Series B 159, 668-686.

Coyette, J. and Ghuysen, J-M. (1968). Structure of the cell wall of *Staphylococcus aureus*, IX. Teichoic acid and phage adsorption. *Biochemistry* 7, 2385-2389.

Derjaguin, B.V. and Landau, L. (1941). Theory of the stability of strongly charged lyophobic sols and of the adhesion of strongly charged particles in solutions of electrolytes. *Acta Physicochimica URSS* 14, 633-662.

Duguid, J.P. and Gillies, R.R. (1957). Fimbriae and adhesive properties in dysentery bacilli. *Journal of Pathology and Bacteriology* 74, 397-411.

Duguid, J.P. and Gillies, R.R. (1958). Fimbriae and haemagglutinating activity in *Salmonella*, *Klebsiella*, *Proteus* and *Chromobacterium*. *Journal of Pathology and Bacteriology*

75, 519.

Duguid, J.P., Smith, I.W., Dempster, G. and Edmunds, P.N. (1955). Non-flagellar filamentous appendages (fimbriae) and haemagglutinating activity in *Bacterium coli*. *Journal of Pathology and Bacteriology* 70, 335-348.

Fletcher, M. (1976). The effects of proteins on bacterial attachment to polystyrene. *Journal of General Microbiology* 94, 400-404.

Fletcher, M. and Floodgate, G.D. (1973). An electron-microscope demonstration of an acidic polysaccharide involved in the adhesion of a marine bacterium to solid surfaces. *Journal of General Microbiology* 74, 325-334.

Garrett, A.J. (1965). Rapid extraction of phosphorus during periodate oxidation of isolated cell walls of *Staphylococcus aureus* (Oxford). *Biochemical Journal* 95 6c.

Gibbons, R.J. (1977). Adherence of bacteria to host tissue. In *Microbiology - 1977*, pp.395-406. Edited by D. Schlessinger. Washington : American Society for Microbiology.

Gibbons, R.J. and Houte, J. van (1975). Bacterial adherence in oral microbial ecology. *Annual Review of Microbiology* 29, 19-44.

Greig, R.G. and Jones, M.N. (1976). The possible role of steric forces in cellular cohesion. *Journal of Theoretical Biology* 63: 405-419.

Hill, M.J., James, A.M. and Maxted, W.R. (1963). Some physical investigations of the behaviour of bacterial surfaces. VIII. Studies on the capsular material of *Streptococcus pyogenes*. *Biochimica et Biophysica Acta* 63, 264-274.

Hogg, R., Healy, T.M. and Fuerstenau, D.W. (1966). Mutual coagulation of colloidal dispersion. *Transactions of the Faraday Society* 62, 1638.

Howard, L. and Tipper, D.J. (1973). A polypeptide bacteriophage receptor: modified cell wall protein subunits in bacteriophage-resistant mutants of *Bacillus sphaericus* Strain P-1. *Journal of Bacteriology* 113, 1491-1504.

Hughes, R.C. and Stokes, E. (1971). Cell wall growth in *Bacillus licheniformis* followed by immunofluorescence with mucopeptide-specific antiserum. *Journal of Bacteriology* 106: 694-696.

Hughes, R.C., Thurman, P.F. and Salaman, M.R. (1971). Antigenic properties of *Bacillus licheniformis*. Cell wall components. *European Journal of Biochemistry* 19, 1-8.

Israelachvili, J.N. and Tabor, D. (1973). In *Progress in Surface and Membrane Science*, vol. 7, pp. 1-55. Edited by J.F. Danielli, M.D. Rosenberg and D.A. Cadenhead. London : Academic Press.

Kelstrup, J. and Funder-Nielson, T.D. (1974). Adhesion of dextran to *Streptococcus mutans*. *Journal of General Microbiology* 81: 485-489.

Kennedy, J.F., Barker, S.A. and Humphreys, J.D. (1976). Microbial cells living immobilised on metal hydroxides. *Nature, London* 261, 242.

Lindberg, A.A. (1973). Bacteriophage receptors. *Annual Review of Microbiology* 27: 205-241.

Makela, P.H. and Mayer, H. (1976). Enterobacterial common antigens. *Bacteriological Reviews* 40: 591-632.

Marshall, K.C., Stout, R. and Mitchell, R. (1971a). Selective sorption of bacteria from sea water. *Canadian Journal of Microbiology* 17: 1413-1416.

Marshall, K.C., Stout, R. and Mitchell, R. (1971b). Mechanism of the initial events on the sorption of marine bacteria to surfaces. *Journal of General Microbiology* 68: 337-348.

Morris, E.R., Rees, D.A., Young, G., Walkinshaw, M.D. and Darke, A. (1977). Order-disorder transition for a bacterial polysaccharide in solution - a role for polysaccharide conformation in recognition between *Xanthomonas* and its plant host. *Journal of Molecular Biology* 110: 1-16.

Mukasa, H. and Slade, H.D. (1973). Mechanism of adherence of *Streptococcus mutans* to smooth surfaces. *Infection and Immunity* 8: 555-562.

Ofek, I., Beachey, E.H., Jefferson, E.H. and Campbell, G.L. (1975). Cell membrane-binding properties of Group A streptococcal lipoteichoic acid. *Journal of Experimental Medicine* 141: 990-1003.

Ofek, I., Mirelman, D. and Sharon, N. (1977). Adherence of *E. coli* to human mucosal cells mediated by mannose receptors. *Nature, London* 365: 623-625.

Ottow, J.C.G. (1975). Ecology, physiology and genetics of fimriae and pili. *Annual Review of Microbiology* 29: 79-108.

Pangborn, J., Kuhn, D.A. and Woods, J.R. (1977). Dorsal-ventral differentiation in *Simonsiella* and other aspects of its morphology and ultrastructure. *Archives of Microbiology* 113: 197-204.

Pethica, B.A. (1961). The physical chemistry of cell adhesion. *Experimental Cell Research* Suppl. 8, 123-140.

Ridgway, H.F. (1977). Ultra-structural characterisation of goblet-shaped particles from the cell wall of *Flexibacter polymorphus*. *Canadian Journal of Microbiology* 23: 1201-1213.

Ridgway, H.F. and Levin, R.A. (1973). Goblet-shaped sub-units from the wall of a marine gliding microbe. *Journal of General Microbiology* 79: 119-128.

Rogers, H.J. (1963). The bacterial cell wall. The result of adsorption, structure or selective permeability? *Journal of General Microbiology* 32: 19-24.

Savage, D.C. (1977). Electron microscopy of bacteria adherent to epithelia in the murine intestinal canal. In *Microbiology - 1977*, pp.422-426. Edited by D. Schlessinger. Washington : American Society for Microbiology.

Scherrer, R. and Gerhadt, P. (1971). Molecular sieving by the *Bacillus megaterium* cell wall and protoplast. *Journal of Bacteriology* 107: 718-735.

Shands, J.W. (1966). Localisation of somatic antigen on Gram-negative bacteria using ferritin antibody conjugates. *Annals of the New York Academy of Sciences* 133, 292-298.

Shattock, P.M.F. and Smith, D.G. (1963). The location of the Group D antigen of *Streptococcus faecalis* var. *liquefaciens*. *Journal of General Microbiology* 31, iv.

Slade, H.D. (1977). Cell surface antigenic polymers of *Streptococcus mutans* and their role in adherence of microorganisms *in vitro*. In *Microbiology - 1977*, pp.411-416. Edited by D. Schlessinger, Washington : American Society for Microbiology.

Smith, H. (1977). Microbial surfaces in relation to pathogenicity. *Bacteriological Reviews* 41: 475-500.

Spinell, D.M. and Gibbons, R.J. (1974). Influence of culture medium on the glucosyl transferase and dextran binding capacity of *Streptococcus mutans* 6715 cells. *Infection and Immunity* 10: 1448-1451.

Staley, T.E., Jones, E.W. and Corley, L.D. (1969). Attachment and penetration of *Escherichia coli* into intestinal epithelium of the ileum in new born pigs. *American Journal of Pathology* 56: 371-392.

Swanson, J. (1975). Role of pili in interactions between *Neisseria gonorrhoeae* and eukaryote cells *in vitro*. In *Microbiology - 1975*, pp. 124-126. Edited by D. Schlessinger. Washington : American Society for Microbiology.

Swanson, J. (1977). Adherence of Gonococci. In *Microbiology - 1977*, pp. 427-430. Edited by D. Schlessinger. Washington : American Society for Microbiology.

Swanson, J., Hsu, K.C. and Gotschlich, E.C. (1969). Electron microscopic studies on streptococci. I.M. antigen. *Journal of Experimental Medicine* 130: 1063.

Takeuchi, A. (1967). Electron microscopic studies of experimental *Salmonella* infections. I. Penetration into the intestinal epithelium by *Salmonella typhimurium*. *American Journal of Pathology* 50: 109-117.

van Driel, D., Wicken, A.J., Dickson, M.R. and Knox, K.W. (1973). Cellular location of the lipoteichoic acids of *Lactobacillus fermenti* NCTC 6991 and *Lactobacillus casei* NCTC 6375. *Journal of Ultrastructure Research* 43: 483-497.

Verwey, E.J.W. and Overbeek, J. Th. G. (1948). *Theory of the*

Stability of Lyophobic Colloids. Amsterdam : Elsevier Publishing Co.

Wagner, R.C. and Barnett, H.R.J. (1974). The fine structure of prokaryotic-eukaryotic cell junction. *Journal of Ultrastructure Research* 48: 483-497.

Weiss, L. (1968). Studies on cellular adhesion. An experimental and theoretical approach to interaction forces between cells and glass. *Experimental Cell Research* 53: 603-614.

Weiss, L. and Harlos, J.P. (1972). Some speculations on the rate of adhesion of cells to coverslips. *Journal of Theoretical Biology* 37: 169-179.

Weiss, L. and Harlos, J.P. (1977). Cell contact phenomena and their implication in cell communication. In *Intercellular Communications*, pp. 33-59. Edited by W.C. de Mello. New York : Plenum Publishing Corporation.

Weiss, L., Nir, S., Harlos, J.P. and Subjeck, J.R. (1975). Long distance interactions between Ehrlich ascites tumour cells. *Journal of Theoretical Biology* 51: 439-454.

Wicken, A.J., Elliot, S.D. and Baddiley, J. (1963). The identity of streptococcal Group D antigen with teichoic acid. *Journal of General Microbiology* 31: 231-239.

Wolpert, J.S. and Albersheim, P. (1976). Host-symbiont interactions. I. The lectins of legumes interact with the O-antigen-containing lipopolysaccharides of their symbiont rhizobia. *Biochemical and Biophysical Research Communications* 70: 729-739.

ADHESION OF MICROORGANISMS IN FERMENTATION PROCESSES

S.G. ASH

Shell Research Limited, Shell Biosciences Laboratory, Sittingbourne Research Centre, Sittingbourne, Kent ME9 8AG.

INTRODUCTION

All fermentation processes, that is the wide range of industrial processes that utilise the properties of microorganisms, involve a large area of interface. The properties of these interfaces and, in particular, their ability to adsorb microorganisms are of major importance in fermentation processes (Table 1). For example, in the manufacture of single cell protein from methane or methanol, 1 m^3 of a typical broth will contain 1 km^2 of cell surface area and approximately 1000 m^2 of gas/liquid interface. In these single-cell-protein processes, microorganisms are grown discretely in deep culture, but many processes use films of microorganisms attached to an inert support material.

An example is the trickle (or percolating) filter used for water purification. In this film reactor the specific surface areas of the inert support (for example, gravel or plastic packing) and of the supported microbial film are made as large as possible within the constraints of practical engineering and economics. As the advantages of retaining microorganisms in the reactor become exploited (Table 2), it will become increasingly important to control the adhesion of cells to support materials and to themselves in the formation of thick microbial films. In the leaching of metal ores and in the digestion of cellulosic materials, the solid substrate presents a large surface area for microbial attachment and attack. Not all microbial films are welcome. The formation of fouling films on heat exchangers and in recovery equipment downstream of fermenters is undesirable, and the presence of films in laboratory scale experimental reactors cannot be disregarded [Munsen and Bridges, 1964; Topiwala and Hamer, 1971].

Table 1.

Adhesion of microorganisms in fermentation processes

Surface	Process/consequence
Reactor walls	Wall growth - fouling of heat exchangers
	- mitigation of wash-out
	- corrosion
Support material	Film formation in packed beds
	eg. trickle filter
	rotating disc
	(active carbon columns!)
	and in fluidised beds
	Biocatalysis (eg. to ion-exchange column)
Nutrient	Digestion of cellulose
	Food and beverage fermentations
Energy source	Leaching of ores
Microorganisms	Flocculent growth, eg. activated sludge
	Flocculation
	Thick film formation
Filter medium	Non-absolute filtration for purification
Various	Fouling of centrifuges, pipework, etc.
Ion-exchange resins	Separation of microorganisms.

The aggregation of microorganisms into flocs is frequently essential to the functioning of a fermentation process. A principle aim of the activated sludge process for the treatment of waste water is the production of an easily settled, flocculated sludge, and failure in the operation of activated sludge plants arises as frequently from the formation of a

Table 2.

Advantages of retaining cells in reactor vessel in continuous processes

1. Wash-out of cells not possible
2. Cell production and biocatalysis processes may be separated.
3. Easier product recovery
4. Reactor stability to varying substrate feed rate.
5. Plug flow in packed columns (high conversions possible)

Surface environment effects (that might be used to advantage)

1. Diffusion limitations.
2. Nutrient adsorption.
3. Denaturing of proteins.
4. pH shift ($pH_{surface} < pH_{bulk}$ for negative surface).
5. Redox potential change.
6. Adsorption of extracellular enzymes.
7. Protection from shear forces.

poorly flocculating (bulking) sludge as from the poor removal of soluble carbon compounds in the waste water.

In brewing by the batch process, the yeast cells must flocculate at the end of the growth phase to allow easy separation by sedimentation (bottom beers) or flotation (top beers). In continuous brewing in tower fermenters the formation of sedimenting flocs is essential to prevent the yeast washing out [Rainbow, 1970]. A major process cost in the production of bacterial or single-cell protein is the dewatering of the broth. This cost is considerably reduced by incorporating a flocculation process to increase the sedimentation rate of the cells within the centrifuges [Edwards, 1969].

The morphology of moulds depends on the state of aggregation of spores and hyphae. In the production of citric acid by *Aspergillus niger* the pellet form is favoured, while in the production of penicillin by *Penicillium chrysogenum*, mycelia are preferred [Whitaker and Long, 1973; Metz and Kossen, 1977].

Instead of relying on the innate forces of attraction for the adhesion of microorganisms to surfaces, it is possible to bind the cells to surfaces by reaction with added chemicals. Such immobilisation of whole cells to support materials is a technique used in the manufacture of biocatalysts.

This brief introduction has indicated the importance of microbe/solid contacts in a wide range of fermentation processes. Three topics have been chosen for further discussion: the adhesion of microorganisms to ores, the attachment of cells to themselves in film formation and flocculation, and the immobilisation of whole cells.

LEACHING OF ORES

Bacteria are used in a variety of hydrometallurgical processes that involve oxidation reactions. Examples are the heap leaching of low grade sulphide-bearing ores and the oxidation of uranium-bearing minerals. There is interest in extending the principles of microbiological leaching to the *in situ* leaching of fractured mineral formations and to the leaching of ores containing intimately mixed metals which, therefore, are not amenable to conventional flotation methods of separation. The subject has been reviewed recently by Torma [1977] and Kelly [1976].

The microorganisms implicated in these processes are autotrophic aerobic (chemolithotrophic) bacteria which derive their energy from inorganic oxidation reactions and which can utilise carbon dioxide as their main carbon source for biosynthesis. Two species of these chemolithotrophic bacteria most frequently studied are *Thiobacillus ferrooxidans*, which oxidises at least ferrous ions and reduced-valency inorganic sulphur compounds, and *Thiobacillus thiooxidans* which oxidises reduced-valency inorganic sulphur compounds, including elemental sulphur, but which cannot oxidise ferrous ions and insoluble metal sulphides.

Some proposed mechanisms by which microorganisms influence the oxidation of sulphide ores are summarised in Table 3. Not all of these outline mechanisms require attachment of the cells to the substrate surface. For example, *Thiobacillus ferrooxidans* can oxidise ferrous ions to ferric ions in solution. The redox potential of the solution is then sufficiently high that the ferric ions can oxidise the ore.

In many processes the leaching by a ferric ion liquor, regenerated *in situ* by microbial activity, is the dominant mechanism, but certainly in the absence of iron and other transit-

Table 3.

Hydrometallurgical processes catalysed by bacteria

Sulphide ores - overall reaction

$$MS + 2O_2 \rightarrow MSO_4$$

Mechanism 1 (direct, requires attached bacteria)

$$MS_{solid} + 2O_2 \xrightarrow{bacteria} M^{2+} + SO_4{}^{2-}$$

Mechanism 2 (indirect, requires presence of iron)

$$MS + 8Fe^{3+} + 4H_2O \rightarrow M^{2+} + SO_4{}^{2-} + 8Fe^{2+} + 8H^+$$

$$4Fe^{2+} + O_2 + 4H^+ \xrightarrow{bacteria} 4Fe^{3+} + 2H_2O$$

Mechanism 3 (sulphur oxidation, requires attached bacteria).

$$MS_{solid} + \tfrac{1}{2}O_2 + 2H^+ \rightarrow M^{2+} + S^0 + H_2O$$

$$2S^0 + 3O_2 + 2H_2O \xrightarrow{bacteria} 2H_2SO_4$$

Mechanism 4 (corrosion cell, promoted by attached bacteria).

$$MS_{solid} \rightarrow M^{2+} + S^0 + 2e^-$$

$$2e^- + 2H^+ + \tfrac{1}{2}O_2 \rightarrow H_2O$$

ion metal ions, and for the oxidation of elemental suphur by *Thiobacillus thiooxidans*, microbial attachment is a prerequisite for attack of the substrate. In addition, attachment of the bacteria to the ore in heap leaching promotes a high concentration of bacteria in the reaction zone, that is the system behaves as a plug flow film reactor.

It is frequently observed [Starkey, Jones and Frederick, 1966; Cook, 1964; Vogler and Umbreit, 1941] that an inoculum of the chemolithotrophic bacteria added to a slurry of the substrate is rapidly adsorbed by the substrate under quiescent conditions. Conventional shake flask agitation may be sufficiently violent to displace some of the inoculated bacteria from the substrate surface and so to reduce the initial rate of leaching. This suggests that the forces of adhesion initially are relatively weak. Following an initial non-agitated period, the initiation of agitation does not lead to the displacement of cells, suggesting that, once attached by a secondary mechanism, the forces of adhesion are relatively strong.

The addition of surfactants, at concentrations insufficient to affect the viability of the cells, together with the inoculum can increase the initial and subsequent rates of reaction [Starkey *et al.*, 1956; Cook, 1964; Duncan, Trussell and Walden, 1964]. The mode of action is not proven, but increased wetted surface area, improved attachment in agitated systems and improved gas transfer have been suggested. Some authors [Duncan *et al.*, 1964] have reported poorer leaching in the presence of added surfactants. It is known that bacteria produce natural surfactants [Jones and Starkey, 1961] and one of these, phosphatidylinositol, has been shown to enhance leaching rates [Schaeffer and Umbreit, 1963].

The attachment of the chemolithotrophic bacteria to surfaces is not specific to substrates providing energy, and attachment to glass [Weiss, 1973], aluminium, [Dugan and Lundgren, 1964] and iron oxide precipitates [McGoran *et al.*, 1969] has been reported. Using electron microscopy, Murr and Berry [1976; also Berry and Murr, 1975] observed no preferential adhesion of cells of *Thiobacillus ferrooxidans* to specific surface features such as steps or dislocations for both molybdenum sulphide and chalcopyrite ($CuFeS_2$). This suggests that there is little surface migration of cells once they are attached. No pili or other holdfasts were observed.

However, the same authors did observe, after eleven days incubation, a considerable growth of bacteria at chalcopyrite inclusions, accompanied by pitting. While this indicates preferential growth on the inclusion and lack of migration of new cells, it does not indicate the preferential adsorption of cells on the inclusion on initial contact of the ore with the inoculum.

Pitting of sulphur crystals which could be attributed to erosion in the immediate vicinity of adhered cells has been observed for both *Thiobacillus thiooxidans* [Schaeffer, Holbert and Umbreit, 1962] and *Sulfolobus* sp. [Weiss, 1973]. *Sulfolobus* exhibited pili, usually 1-2 μm in length and having a diameter of 5 nm. Good adhesion to sulphur crystals occurred when pili were present, but when the bacteria were grown in the absence of yeast extract the number of pili was greatly reduced and the bacteria were unable to attach to sulphur.

Attachment and growth of bacteria on mineral surfaces is confirmed and evidence suggests a strong interaction, but an understanding of the mechanisms of attachment and of chemical erosion awaits further investigation.

AGGREGATION AND THICK FILM FORMATION

The aggregation of microorganisms and the formation of a thick microbial film are related phenomena which arise from the adhesion of cells to themselves. A number of terms are used to describe the aggregation of cells, including flocculation, flocculence, coagulation, agglutination and clumping. In general these words are used interchangeably, but a mechanistic distinction (to be discussed later) can be drawn between the coming together of once discrete cells (flocculation) and the non-separation of daughter cells from their parents (flocculent growth).

Unfortunately, the definition of an aggregated state of microorganisms may be as arbitrary as the choice of word to describe it, since both process engineers and research workers seeking the ideally flocculated state for their microorganisms may ignore less satisfactory states of aggregation, for example, weak or small flocs, and refer to these states as non-aggregated.

To operate satisfactorily, many fermentation processes require cell-cell adhesion and it is not surprising that the phenomenon has been extensively studied. However, much of the study has been qualitative rather than quantitative and microbiological rather than physical in nature, such that many questions concerning the mechanisms of cell-cell adhesion remain to be answered. Given the right selective pressure, the microorganisms with the desired flocculent properties tend to appear and the colloid scientist has not been in such great demand by biochemical engineers as one might expect from the introductory survey.

The relevance of flocs and of microbial films to biochemical engineering has been the subject of reviews by Atkinson and Daoud [1976] and Atkinson and Fowler [1974]. The aim of this review is to examine the extent to which current theories of colloid stability are able to explain the mutual adhesion or repulsion of some microorganisms of industrial significance.

Nature of forces in colloidal systems

The nature of forces operating in colloidal systems has been reviewed in the context of cell adhesion by several authors [Pethica, 1961; Curtis, 1967; Lips, 1978]. Since the establishment of the DLVO theory [Derjaguin and Landau, 1941; Verwey and Overbeek, 1948] two major advances in colloid science have been the understanding of the nature and magnitude of van der Waals forces between condensed phases [Israelachvili and Ninham, 1977; Israelachvili, 1973; Parsegian, 1973] and of those forces arising from the presence of macromolecules either adsorbed or attached to approaching surfaces [Napper,

1977; Vincent, 1974].

The total interaction between two microorganisms is the summation of a wide range of forces, each contributing to a lesser or greater extent. To ask which force causes flocculation is therefore meaningless. Rather one might ask which forces are dominant. The experimental techniques applied to microorganisms have been barely adequate even to indicate the dominant mechanisms of flocculation and dispersion, and are inadequate to shed any light on lesser forces. The more important forces (in the view of the author) are listed in Table 4. These will be discussed in more detail later, but an immediate explanation of two factors is necessary. The production of extracellular material causes cells trapped in a matrix of that material to move apart against viscous forces. This would account for the wide spacing between cells observed in some microbial films.

Table 4.

Forces between microorganisms in the context of fermentation processes

Factors favouring dispersion
Electrostatic forces (double layer)
Steric forces (attached or adsorbed macromolecules)
Excretion of material between cells (repulsive force)
Hydrodynamic forces (cell motility and shear stresses)
Factors favouring aggregation
Matrix entrapment (viscous effect)
Incipient flocculation (interaction between attached or adsorbed macromolecules involving short range forces including hydrogen bonding, triple ion interactions, double ion interactions)
Charge mosaic interaction (flocculation by cationic polyelectrolytes)
Polymer bridging (flocculation by addition of polymers, also involving short range forces)

Incipient flocculation [Napper, 1977] arises when the solvent conditions are such that the macromolecules adsorbed or attached to the surfaces of the microorganisms tend to phase separate or precipitate. This mechanism is distinguished from the polymer bridging in which the bridging polymers are adsorbed to the opposing surfaces simultaneously.

For a given microbial species, different forces (possibly of the same sign) may not be dominant under different conditions of growth in continuous or batch culture or during different growth phases in batch culture. If a change in conditions gives rise to a change in the state of aggregation of a culture, the cause may be a direct (colloidal) or indirect (physiological) effect. For example, the addition of ions may influence the double layer repulsion or short range interactions between surface groups or, alternatively or simultaneously, the addition of ions may bring about a change in the cell wall structure by a physiological effect. Frequently there are inadequate physical measurements reported to allow a distinction to be made.

It is important to distinguish between flocculent growth and the aggregation of dispersed cells, although Friedman *et al.* [1969] suggest that the two phenomena have the same basis. Two cells aggregate when they have sufficient kinetic energy to overcome any energy barrier as they approach each other from the direction of infinite separation. Two cells remain joined when they have insufficient energy to overcome the energy barrier from the direction of zero separation (Figure 1). Consequently, cells growing as flocs or clumps, if disrupted by high shear conditions, will not necessarily re-aggregate under quiescent conditions. Further, the growth of microorganisms in flocs or as discrete cells yields no information concerning the presence or absence of an energy barrier to flocculation but does indicate the depth of the primary minimum. The fact that so many microorganisms grow singly suggests that the primary minimum is frequently shallow. Thus, to flocculate such cells requires a strengthening of the attractive forces in the primary minimum and, possibly, the weakening of an energy barrier opposing aggregation.

Energy-separation diagrams of the type shown in Figure 1 are gross simplifications. During the brief encounter of two microorganisms, thermodynamic equilibrium is unlikely to exist and the forces will be time-dependent.

The force between the microorganisms is only one of the factors to be considered in the aggregation process. Their collision frequency will influence the kinetics of flocculation. For those of diameter greater than 1 μm, moderate shear stresses in the system such as produced by gentle stirring or convection currents considerably enhance the collision fre-

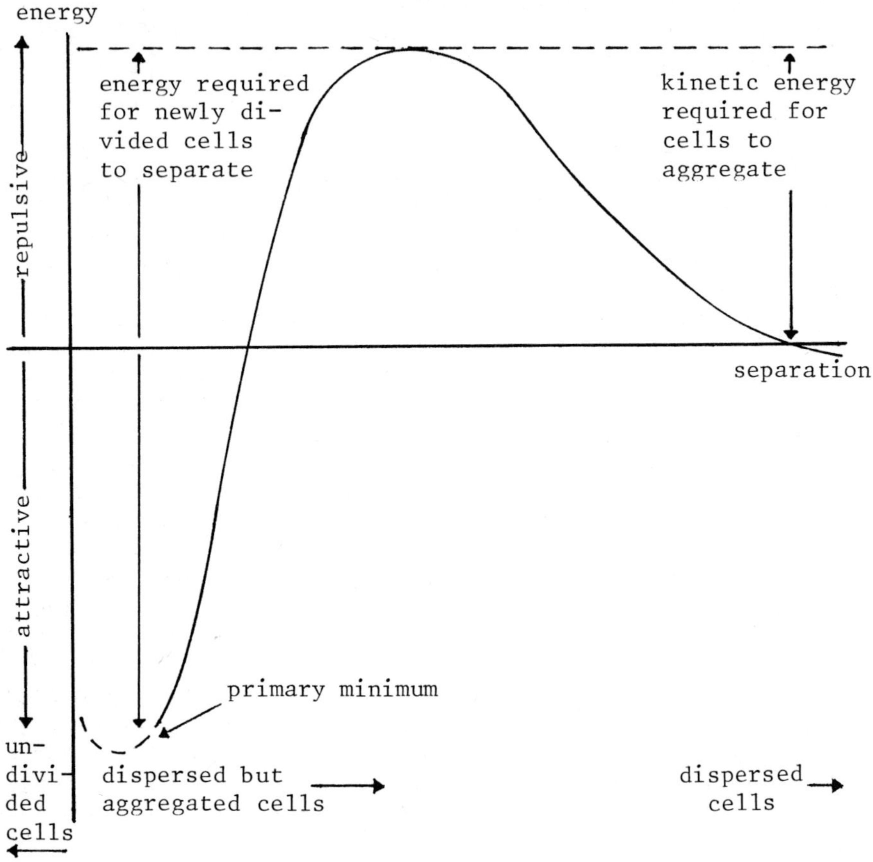

Fig. 1. Energy-separation diagram for two microorganisms showing distinction between aggregation and flocculent growth.

quency. Likewise motility can be important. Thus the use of rates of flocculation, as a guide to the flocculation characteristics of microorganisms, must be measured under defined conditions of temperature, shear, motility and particle number concentration. The collision frequency increases as the square of the concentration of microorganisms.

Finally, the statistical mechanics of aggregation should not be ignored. Long, Osmond and Vincent [1972] have demonstrated that for flocculation into a shallow energy minimum, a dynamic equilibrium exists between flocs and single particles at sufficiently high particle concentrations. Thus the existence of conditions for the flocculation of a dispersion does not necessarily lead to the flocculation of all the particles in that dispersion. The morphology of the flocs will depend on many of the factors discussed above.

Surfaces of microorganisms

While the microbiologist frequently considers the outermost part of a microorganism to be the cell wall, the colloid scientist is interested in the surface presented to the environment. The characteristics of this surface may arise from material excreted by the microorganism or adsorbed from the medium.

There are several bacteria that produce extracellular polysaccharides of commercial interest [Jeanes, 1977]. *Xanthamonas campestris* produces a polysaccharide, with the commercial name Kelzan, which consists of a cellulose backbone with alternate residues attached to side chains containing three residues with anionic character [Janssen, Kenne and Lindberg, 1975]. *Leuconostoc mesenteroides* produces dextran and *Azotobacter* spp. produce microbial alginates. *Zoogloea ramigera*, a frequent component of activated sludge, produces a polysaccharide with β(1-4) linkages [Friedman *et al.*, 1968]. The solubilities of microbial extracellular polysaccharides span a wide range, the least soluble polymers forming capsules about the bacteria.

During the decline phase of growth or under adverse conditions, cells may lyse rapidly and a variety of polymeric materials enters the medium. The aggregation of bacteria during an endogenous growth phase has been attributed to the presence of these products of lysis or to the increased production of extracellular polymers [Tenney and Stumm, 1965; Busch and Stumm, 1968; Cripps and Work, 1967]. For a heterogenous bacterial population, derived from a biotreater, in the endogenous growth phase, Pavoni, Tenney and Echelberger [1972] [see also Tenney and Verhoff, 1973] showed that the extracellular polymeric material comprised polysaccharides, proteins and nucleic acids in the approximate ratio of 3:2:2. Nucleic acids have been implicated in the flocculation of *Pasteurella pestis* [Wessman and Miller, 1966].

Other polymeric materials in the medium may arise from the proteinaceous nutrients used in industrial processes, such as yeast extract and hydrolysed soya. In the biotreatment of domestic effluents or of effluent from the food industry, macromolecules will be present in the waste or may arise from the protozoa, in the biotreater [Parker, Kaufmann and Jenkins, 1971; McKinney, 1956]. The influence of these natural polymers and of synthetic polymers and polyelectrolyte flocculants on the state of aggregation of microorganisms implies that macromolecules generally adsorb readily to the cell surfaces. Likewise, desorption of adsorbed polymers by washing can also significantly alter the surface properties of cells [Gasner and Wang, 1970: Harris and Mitchell, 1975].

Extracellular polymers readily bind ions from solutions [Friedman and Dugan, 1968a] so that the concentration of ions in solution can significantly affect the surface properties of the bacteria. Lower molecular weight organic materials such as lipids and sugars may also adsorb at the cell surface and modify the colloidal properties of microorganisms [Roberts, Wennerberg and Friberg, 1974]. The role of polymers in microbial aggregation has been reviewed by Harris and Mitchell [1973].

The presence and presumed adsorption onto the cells of many ill-defined materials in the medium surrounding microorganisms deters colloid scientists from studying microbial aggregation. At the same time, there is a tendency for microbiologists and biochemists to ignore the impact of these adsorption processes on the stability of dispersions of cells. Yet, provided the environmental conditions during growth are well-defined and controlled, as in continuous culture, and media simplified, in the laboratory at least, cells can be grown with reproducible surface characteristics which may be modified purposefully by changes in the growth conditions [Ellwood and Tempest, 1972; Sutherland, 1975] or by subsequent treatment.

Factors favouring the dispersion of bacteria

Whether microorganisms are regarded as 'hard' bodies with well-defined surfaces or as 'soft' bodies with a diffuse layer of polymeric material, long-range electrostatic interactions may contribute significantly to the total interaction energy between them. However, long-range electrostatic interactions are not necessarily dominant and there are flocculation-deflocculation transitions which do not involve a change in the zeta potential of the cells [McKinney, 1956; Pavoni *et al.*, 1972].

The majority of microorganisms studied have isoelectric points in the range pH 2 - pH 3, reflecting the significant contribution of carboxyl groups to the surface charge at neutral pH values. Electrophoretic mobilities of microorganisms at neutral pH values lie in the range -0.4 to -4.0 (μm/sec)/(volt/cm). At the lower end of this range the resulting electrostatic repulsion is probably small relative to thermal energy, while at the higher end the repulsion will contribute significantly to the total interaction energy between the cells [Wiersema, Loeb and Overbeek, 1966; Pethica, 1961].

Adsorption of material to the surface of cells can alter the electrophoretic mobility of these cells and when excess polymeric material is adsorbed, the cell will assume the electrophoretic characteristics of that material. Indeed, adsorption of macromolecules onto particles which are then examined by microelectrophoresis is an established technique for studying the electrophoretic properties of those macromolecules. The isoelectric point (iep) of many proteins lies in the range pH 4 - pH 6, so that the adsorption of proteins will raise the iep of microorganisms [Northrop and DeKruif, 1922]. At the isoelectric point the adsorption of many proteins will be a maximum [McLaren, 1954; Lyklema and Norde, 1973].

Flocculation of microorganisms in the vicinity of the iep and the formation of stable dispersions at values of pH not near the iep have been observed for a number of systems. Harris and Mitchell [1973] observed that suspensions of glucose-grown *Leuconostoc mesenteroides* aggregated at the iep (pH 3), but were stable at other pH values up to pH 7. These cells produce minimal extracellular dextran. The cause of aggregation above pH 7 was not pursued. For a mixed bacterial population, Pavoni, Tenny and Echelberger [1972] observed a turbidity minimum at pH 2, whilst Forster [1968] showed that the sludge volume index of an activated sludge increased linearly with the magnitude of the electrophoretic mobility. A well-flocculated sludge has a low sludge volume index. The electrophoretic mobility of this mixed microbial population became more negative as the ratio of nitrogen to phosphorus in the growth medium decreased. Northrop and DeKruif [1922] observed the flocculation of *Bacillus* sp. in the vicinity of zero mobility, brought about by the addition of ions or pH adjustment.

Although electrophoretic mobilities were not reported, the maximal clumping of *Corynebacterium xerosis* at pH 3.0 observed by Stanley and Rose [1967] and the flocculation of a suspension of *Flavobacterium* sp. by the addition of monovalent ions [Tezuka, 1969] also suggest that the stability of dispersions of these bacteria depend on double layer repulsion.

The magnitude of wall-growth frequently depends on the pH in the fermenter. Wilkinson and Hamer [1974] observed a thick wall-growth on glass at a pH greater than 5.8 when growing a defined mixed culture on methane. At pH values below 5.7 the wall growth dropped off. Atkinson and Fowler [1974] observed extensive wall-growth for a mixed culture growing at pH 7, much less growth at pH 9 and no thick film formation at pH 5. However, such results cannot be related in any simple way to changes in the double layer repulsion with changes in the hydrogen ion concentration.

The use of natural hydrophilic colloids such as starch or gelatin to stabilise aqueous dispersions has been recognised for many years [Vincent, 1974]. The protective mechanism is now called steric stabilisation, and arises from the overlapping of diffuse layers of adsorbed or attached polymers under conditions such that the overlapping of the polymer layer is unfavourable in terms of the free energy of the system. This unfavourable interaction can arise from a loss of configurational entropy of the polymers but may also be the consequence of an increase in enthalpy due to the release of bound water molecules [Napper and Hunter, 1972]. Because the cores of the particles are kept apart by the diffuse polymer layers, the van der Waals attractive forces are low and easily overcome by thermal energy. Consequently the particles do not aggregate. Evidence for this mechanism operating with microorganisms comes from the failure of some microbes to coagulate at their isoelectric point or in the presence of excess monovalent ions when double layer repulsion is negligible, together with the observation of polymeric material at the surface.

Harris and Mitchell [1975] grew *Leuconostoc mesenteroides* on either sucrose or glucose as the carbon source, and studied the aggregation of the cells as a function of growth conditions and the hydrogen ion concentration, adjusted after growth had finished. Sucrose-grown cells synthesise extracellular dextran but those grown on glucose do not produce dextran. The sucrose-grown cells were completely stable over a wide range of pH values while those grown on glucose aggregated at their isoelectric point and above pH 7. This suggested that the dextran inhibited aggregation of the sucrose-grown cells. This hypothesis was confirmed by centrifuging and resuspending the sucrose-grown cells to remove dextran, whereupon the cells behaved colloidally as if they were glucose-grown cells. Addition of the dextran to resuspended sucrose-grown cells or to the glucose-grown cells resulted again in stable dispersions.

Excess dextran can stabilise suspensions of *Escherichia coli* and *Aerobacter aerogenes* [Busch and Stumm, 1968] and dispersions of several non-biological colloids [Tenney and Stumm, 1965; Bontaux, Dauplan and Marignan, 1969].

McGregor and Finn [1969] observed that *Pseudomonas fluorescens* which liberated RNA into the medium, did not flocculate when the electrophoretic mobility was reduced to zero by the addition of 0.02 M sodium acetate.

Factors favouring the aggregation of bacteria

Electron micrographs and photographs of flocs from activated sludge [Friedman and Dugan, 1968b; Friedman, *et al.*, 1968; Friedman *et al.*, 1969; Crabtree *et al.*, 1966; Mueller, Morand and Boyle, 1967; Warren and Gray, 1955; Finstein, 1967; Tezuka, 1967; Kiuchi *et al.*, 1968] and of films [Unz and Dondero, 1967] show that the cells are frequently surrounded by a polymeric material, usually identified as a polysaccharide. Characklis [1977] has measured a significant elastic modulus and a very high viscous modulus for a typical biofouling film which behaved as if it were a lightly cross-linked gel. The rheological properties of extracted microbial polysaccharides have been extensively studied because of their industrial applications as thickening additives [Jeanes, 1977]. In general their high molecular weights (frequently 10^6 - 10^8) and molecular rigidities lead to very viscous solutions at moderate concentrations, and some are capable of forming gels under certain chemical conditions [Rees and Welsh, 1977].

It is proposed that one mechanism leading to the formation of flocs or film is the entrapment of microorganisms in an extracellular polymeric gell or viscous matrix. This mechanism is independent of any thermodynamic considerations which may suggest the stable state of the system is dispersion rather than aggregation, for example, by the mechanism of steric stabilisation. The high viscosity of the matrix considerably reduces the rate at which such a stable dispersion could be formed.

The observation that cells of *Zoogloea ramigera* reaggregate under quiescent conditions following mechanical dispersion [Crabtree *et al.*, 1966] indicates that in this system aggregation is also the thermodynamically stable state.

The excretion of large quantities of polysaccharide *per se* is insufficient to cause flocculent growth or the formation of thick films, since *Xanthomonas campestris* grows as discrete cells under conditions of mild agitation. On the other hand, cells which do not excrete polysaccharides may be trapped in the polymeric matrix produced by other bacteria in a heterogenous population such as activated sludge [Crabtree, 1971]. The likelihood of cells becoming trapped in a polymeric matrix must depend on the relative rates of production of extracellular polymer and dissolution of the polymer into the bulk medium away from the cell surface. Dissolution will be greatest in regions of highest agitation and this may account for the formation of films, preferentially in regions of low shear, in processing equipment. At the same time, the excretion of polymeric material into the space between trapped cells could lead to a force of repulsion between the cells and give rise to the wide spacing between cells sometimes observed in films.

Steric stabilisation of particles by adsorbed or attached macromolecules arises because overlapping of the polymer layers is unfavourable and would increase the free energy of the system. However, if the free energy change upon overlapping the polymer layers is negative, then incipient flocculation is the result [Napper, 1977; Vincent, 1974]. Napper and Hunter [1972] observed for several synthetic polymers stabilising dispersions of polymer lattices that flocculation occurred when the aqueous phase approached θ solvency conditions, either by the addition of electrolytes or by changing the temperature to the consulate point.

In summary, those conditions likely to bring about precipitation or phase separation of the polymers in bulk solution are the same conditions expected to cause the flocculation of a dispersion protected by an adsorbed or attached layer of that polymer. For a discussion of the details see Vincent [1974]. The author knows of no comprehensive studies correlating the solution thermodynamic properties of biopolymers and the flocculation behaviour of microorganisms known to be coated with these biopolymers. Such studies would greatly clarify the role of incipient flocculation in the aggregation of microorganisms.

Conditions for precipitating biopolymers from solutions for purification and concentration are well-established. Thus proteins may be precipitated by adjusting the solution pH to the isoelectric point, adding electrolyte and cooling (the salting out effect). Polysaccharides are precipitated by adding electrolytes and a non-solvent miscible with water such as methanol, or by adding a quaternary ammonium compound. It is to be expected, therefore, that such conditions will lead to the flocculation of microorganisms coated by such biopolymers.

Alternatively, incipient flocculation may be viewed in terms of the short-range bonding between polymer segments. Hydrogen bonding, ion pair bonding (for example, $-NH_4^+\ ^-OOC-$) and triple ion bonding involving polyvalent cations (for example, $-COO^-Ca^{2+}\ ^-OOC-$) have been proposed as the most important of the short-range interactions in biopolymers [Pethica, 1961]. Those conditions enhancing the number or strength of these bonds will promote flocculation.

The important role of divalent cations, in particular calcium, in the flocculation of bacteria has been postulated or demonstrated by a number of authors [Friedman *et al.*, 1969; Tenney and Stumm, 1965; Stanley and Rose, 1967; Tezuka, 1969; Hodge and Metcalf, 1958]. However, the concentration of magnesium ions in the medium during growth can influence the cell wall composition of bacteria [Ellwood and Tempest, 1972] and the suppression of flocculation by magnesium ions present during growth has been observed [Tezuka, 1967].

The conditions likely to bring about incipient flocculation, that is a pH value near the isoelectric point and a high ionic strength, are similar to those which reduce double layer repulsion. Hence, from observations of the influence of the hydrogen ion concentration and ionic strength alone on the stability of microorganisms, it is difficult to distinguish between the two mechanisms of flocculation, if indeed they are not both operating at the same time.

Incipient flocculation and polymer bridging are two distinct mechanisms (Figure 2). Polymer bridging is defined as the adsorption of a polymer molecule onto more than one part-

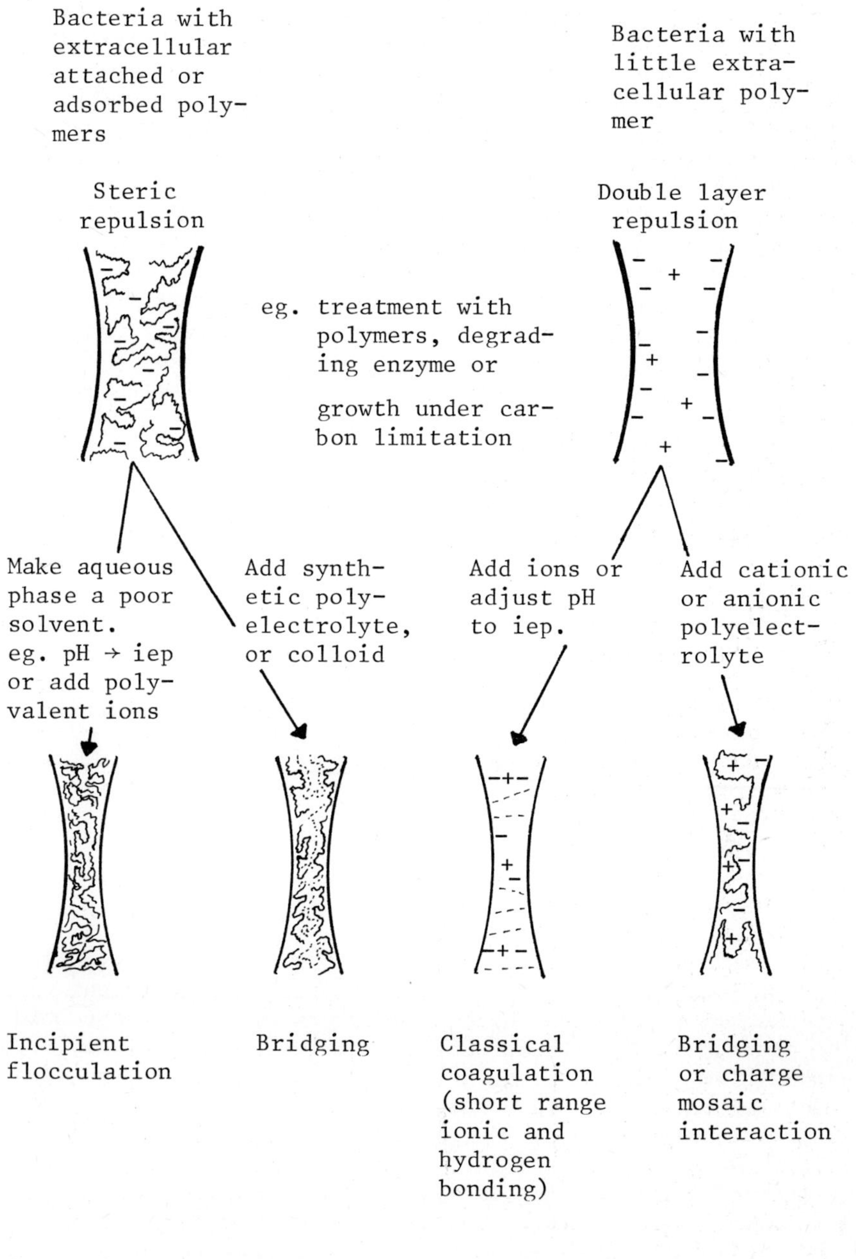

Fig. 2. Incipient flocculation and polymer bridging.

icle surface at the same time. Indeed, the strong interaction between polymer molecules required for incipient flocculation is undesirable for flocculation by the bridging mechanism since an extended configuration of the molecule favours the formation of polymer bridges. In the absence of quantitative measurements of the state of the polymers at the surface of microorganisms, it is difficult to distinguish between the mechanisms of incipient flocculation and polymer bridging, and many examples of polymer bridging quoted by authors may, in fact, be examples of incipient flocculation.

Crabtree *et al.*, [1966] were unable to grow a strain of *Zoogloea ramigera* in a dispersed state when excess nitrogen was present in the medium. The addition of excess carbon source to a growing culture led to the production of a storage carbohydrate and flocculation within thirty minutes. Deinema [1972] showed that flocculent cells of the same strain were surrounded by cellulose. Degradation of the cellulose surrounding this strain and many other strains of bacteria by cellulase dispersed the cells [Deinema and Zevenhuizen, 1971]. It might be expected that such an insoluble polysaccharide would lead to flocculent growth by an incipient flocculation mechanism.

Under usual conditions of growth, *Nocardia corallina* forms large clumps that remain after suspension and agitation in buffer solutions at pH values from 2 to 9 [Clark, 1958]. Electron micrographs revealed a slime layer connecting the cells. This material can be washed off by distilled water, such that the tendency for the cells to aggregate is reduced. However, the material could not be removed by washing in various salt solutions, dilute acids or bases, or organic solvent, suggesting that the material is insoluble in these liquids and lending support to an incipient flocculation mechanism under normal environmental conditions.

Polymer bridging may arise when a polymer is addded to a suspension of microorganisms. When the microorganisms are already coated with biopolymer, the bridging flocculant will need to interact with these biopolymers by short-range bonds as previously described. Cationic polyelectrolytes may bring about flocculation of the negatively charged microorganisms by reducing the net surface charge or by introducing interactions between charge mosaics [Gregory, 1973; Treweek and Morgan, 19 77]. The extent of flocculation will pass through a maximum with increasing polymer dosage, since excess polymer leads to steric or electrostatic repulsion. Such maxima are common when cells are flocculated by synthetic polyelectrolytes and optimal concentrations of added biopolymers have also been observed [Busch and Stumm, 1968; Tenney and Stumm, 1965; Hodge and Metcalf, 1958].

In a heterogenous population, heteroflocculation between cells with surface polymers and those without could occur. Heteroflocculation between cells and non-biological particles has been observed [Harris and Mitchell, 1974], and it is a general observation that mixed microbial populations form flocs and wall growth more readily than mono-cultures. Kiuchi *et al.* [1968] isolated the twelve major component strains from an activated sludge and all strains formed no flocs or only weak flocs when grown in monoculture. However, strong flocs formed when the cells were grown together.

Brewer's yeast

The flocculation behaviour of *Saccharomyces cerevisiae* which aggregates about carbon dioxide bubbles to form a yeasty head, and of *Saccharomyces carlsbergensis*, which aggregates and sediments, has been extensively studied. In brewing, the maintenance of a yeast with the desired flocculant characteristics is a refined art since these characteristics are not stable [Rainbow, 1970] and yeasts may develop the tendency to flocculate too early in the batch brewing process or not at all.

The existence of flocculating and non-flocculating strains of yeast had led to a search for the difference in structure between the two groups of strains. Naturally, the cell wall has been the focus of attention. It is important to note that the terms flocculent and non-flocculent in the context of brewer's yeast often refer to the state of aggregation in that environment existing at the end of a batch fermentation. In a different environment, flocculent yeasts may remain dispersed and non-flocculent yeasts may aggregate. The focus of attention on yeast cell walls was justified when it was demonstrated that separated cell walls behaved colloidally like the original cells [Eddy and Rudin, 1958a; Maaschelein and Devreux, 1957].

The outer cell wall of *Saccharomyces cerevisiae* [Matile, Moor and Robinow, 1969] comprises a protein/mannan complex containing hydroxy, carboxy and phosphate groups, responsible for the observed net negative charge at values of pH greater than 2.3 [Eddy and Rudin, 1958b]. Beers and worts have an acidity in the range pH 3.8 to pH 5.6 [Rainbow, 1966]. The pK_a of the yeast surface is 5, and the ionic spectrum suggests a predominance of carboxyl groups in the surface [Jensen and Mendlik, 1951]. Adsorbed on the cell wall may be proteins and other molecules present in the wort.

Multivalent ions are required for flocculation since washing a suspension of cells with distilled water destroys the flocs which reform on addition of multivalent ions [Jensen and Mendlik, 1951]. At moderate concentrations, monovalent ions have no influence. Calcium ions are particularly active [Jansen and Mendlik, 1951; Eddy, 1955a; Mill, 1964] but other multivalent ions can induce flocculation [Lindquist, 1953; Eddy, 1955c]. Complexing calcium ions in the medium using EDTA leads to deflocculation [Taylor and Orton, 1973]. These observations together with the knowledge of the cell wall structure suggest that polyvalent ion bridges are the dominant interaction in the attraction between flocculated yeast cells. Other forces of attraction, such as hydrogen bonding, may also operate [Mill, 1964]. The adsorption of calcium ions by yeast is a maximum at pH 4.5 to pH 5.5, corresponding to the optimum pH for flocculation [Mill, 1964] and the pK_a for carboxyl groups.

The mechanism of calcium ion bridging has been proposed by Gasner and Wang [1970] for the flocculation of yeasts *(Candida intermedia)* by strongly anionic polymers and similar observations have been made in physical systems [Sarkar and Teot, 1973].

However, the presence of calcium ions does not necessarily lead to the flocculation of brewer's yeast, for example, at the start of a batch fermentation process [Geilenkotten and Nyns, 1971]. A wide range of added materials is found to inhibit flocculation including carbohydrates [Lindquist, 1953; Eddy, 1955a; 1955b] and proteins [Baker and Kirsop, 1972]. Jensen and Mendlik [1951] observed no inhibition of flocculation by proteins derived from wort however, and finings, that is proteins near their isoelectric point, are used to flocculate yeasts. It has been proposed that the adsorption of organic molecules prevents the formation of calcium bridges [Lindquist, 1953; Mill, 1964].

An alternative explanation of the stability of some yeast suspensions in the presence of calcium lies in variations in the cell wall structure. The aggregation of flocculent yeasts is suppressed by enzymatic degradation of the cell wall protein using papain [Eddy and Rudin 1958a] or trypsin [Lyons and Hough, 1971] or by inhibiting mannan synthesis using actidione [Baker and Kirsop, 1972]. Mill [1966] found a correlation between the quantity of phosphorylated mannan in the cell walls of a number of yeast strains and the flocculent behaviour of those strains.

There is little correlation between the electrophoretic mobility of *Saccharomyces* and its state of aggregation [Jansen and Mendlik, 1951; Morris, 1966], and the cells do not readily flocculate at their isoelectric point. However, weak attraction may be easily overcome by shear forces for such large cells, and weak flocculation may have been ignored in comparison to the strong flocculation brought about by calcium bridging.

Thus, the evidence suggests that calcium bridges contribute

an attractive force between yeast cells. This force is not always sufficient to aggregate the cells, either because of insufficient bridges (too shallow an energy minimum) or because the forces of repulsion arising from electrostatic interactions or from the presence of adsorbed material forming a steric barrier prevent the cells from approaching sufficiently closely to enable bridging to occur (too high an energy barrier).

Conclusions

Despite the volume of literature on the subject of microbial aggregation, the identification of mechanisms in the light of current knowledge in colloid science is difficult because inadequate information is reported. Thus it is not easy to distinguish between long range electrostatic repulsion and steric hindrance, since both mechanisms can show the same dependence on pH and ionic strength. Likewise it is difficult to distinguish between polymer bridging and incipient flocculation mechanisms of aggregation by biopolymers produced *in situ*. These mechanisms can be resolved only by parallel examination of the thermodynamic properties of the biopolymers in solution or in the adsorbed state. In the latter case, the behaviour of model, well defined colloids, with adsorbed biopolymers may prove useful. Nevertheless, the observations do suggest that a range of mechanisms is operating and form a basis for directing further research.

IMMOBILISATION OF MICROORGANISMS

In the film fermenter (for example, a trickle filter) the cells become attached to the support with little encouragement by the scientist or engineer apart from the imposition of a selection pressure for adhesion during growth and a sensible choice of solid support material. Unfortunately, the extent, stability and activity of the films are frequently ill-defined.

A better defined and more flexible system can be made by growing the microorganisms in deep-culture, recovering and treating the cells to block unwanted pathways or to increase the permeability to products, for example, and finally attaching the cells by chemical means to a support material. These immobilised cells can then be placed in a reactor and used to catalyse the desired reaction. The feedstock for the reaction need not be the substrate required for growth. Indeed, the cells are frequently not in a condition fit for growth and growth in the reactor is not desirable.

Immobilisation of whole cells is a process under development, but a number of means have been used with some success

Table 5.

Means of immobilising microorganisms

1. Entrapment in polymeric gel or inorganic matrix [Jack and Zajic, 1977].
2. Adsorption on solids, e.g.
   - On ion exchange resins [Hattori, Hattori and Furusake, 1972].
   - On transition metal hydroxides [Kennedy, Barker and Humphreys, 1976].
   - On adsorbed layers of protein or polysaccharide [Harisberger, 1976; Griffiths and Compere, 1976].
3. Covalent or co-ordinate bonding:
   - eg. carbodiimide - agarose adipic hydrazide beads [Jack and Zajic, 1977].
   - glutaraldehyde [Martin and Perlman, 1976; Navarro and Durand, 1977].
4. Flocculation by polymers [Lee and Long, 1974].

(Table 5, and see Jack and Zajic, 1977). In general, these techniques are extensions of those applied to the immobilisation of enzyme molecules. The chemist has added his adhesives to those supplied by nature.

CONCLUSION

Biochemical engineers recognise the practical importance of the adhesion of microorganisms and have exploited the phenomemon without the need to understand it. Future developments in biochemical engineering, in particular improvements in reactor design, an increase in production scale and a broadening of the range of products will undoubtedly lead to a further exploitation and this may provide an impetus for more basic studies of the mechanisms of adhesion of microorganisms, of commercial interest, to solid surfaces.

I should like to express my appreciation to Dr. G. Hamer and Dr. B. Vincent for helpful and valuable suggestions.

REFERENCES

Atkinson, B. and Daoud, I.S. (1976). Microbial flocs and flocculation in fermentation process engineering. *Advances in Biochemical Engineering* 4, 41-124.

Atkinson, B. and Fowler, M.W. (1974). The significance of microbial film in fermenters. *Advances in Biochemical Engineering* 3, 221-227.

Baker, D.A. and Kirsop, B.H. (1972). Flocculation in *Saccharomyces cerevisiae* as influenced by wort composition and by actidione. *Journal of the Institute of Brewing* 78, 454-458.

Berry, V.J. and Murr, L.E. (1975). Bacterial attachment to molybdenite: An electron microscope study. *Metallurgical Transactions* 6B, 488-490.

Bontaux, J., Dauplan, A. and Marignan, R. (1969). Stabilisation des colloides mineraux. *Journal de chimie physique et de physiochimie biologique* 66, 1259-1263.

Busch, P.L. and Stumm, W. (1968). Chemical interactions in the aggregation of bacteria. *Environmental Science and Technology* 2, 49-54.

Characklis, W.G. (1977). Biofouling film development and destruction: Experimental systems: *Microbiology of Power Plant Thermal Effluents Symposium, Iowa.*

Clark, J.B. (1958). Slime as a possible factor in cell clumping in *Nocardia corallania. Journal of Bacteriology* 75, 400-402.

Cook, T.M. (1964). Growth of *Thiobacillus thiooxidans* in shaken culture. *Journal of Bacteriology* 88, 620-623.

Crabtree, K., Boyle, W., McCor, E. and Rohlich, G.A. (1966). A mechanism of floc formation by *Zoogloea ramigera. Journal Water Pollution Control Federation* 38, 1968-1980.

Cripps, R.E. and Work, E. (1967). The accumulation of extracellular macromolecules by *Staphylococcus aureus* grown in the presence of sodium chloride and glucose. *Journal of General Microbiology* 49, 127-137.

Curtis, A.S.G. (1967). *The Cell Surface: Its Molecular Role in Morphogenesis.* London : Academic Press.

Deinema, M.H. (1972). Bacterial flocculation and production of poly-β-hydroxybutyrate. *Applied Microbiology* 24, 857-858.

Deinema, M.H. and Zevenhuizen, L.P.T.M. (1971). Formation of cellulose fibrils by gram negative bacteria and their role in bacterial flocculation. *Archiv fur Mikrobiologie* 78, 42-57.

Derjaguin, B.V. and Landau, L. (1941). Theory of the stability of strongly charged lyophobic sols and of the adhesion of strongly charged particles in electrolytic solutions.

Acta Physiochimica URSS 14, 633-662.
Dugan, P. and Lundgren, D.G. (1964). Acid production by *Ferrobacillus ferrooxidans* and its relation to water pollution. *Development in Industrial Microbiology* 5, 250-257.
Duncan, D.W., Trussell, P.C. and Walden, C.C. (1964). Leaching of chalcopyrite with *Thiobacillus ferrooxidans:* Effect of surfactants and shaking. *Applied Microbiology* 12, 122-126.
Eddy, A.A. (1955a). Flocculation characteristics of yeasts. I. Comparative survey of various strains of *Saccharomyces cerevisiae*. *Journal of the Institute of Brewing* 61, 307-312.
Eddy, A.A. (1955b). Flocculation characteristics of yeasts. II. Sugars as dispersing agents. *Journal of the Institute of Brewing* 61, 313-317.
Eddy, A.A. (1955c). Flocculation characteristics of yeasts. III. General role of flocculating agents and special characteristics at a yeast flocculated by alcohol. *Journal of the Institute of Brewing* 61, 318-320.
Eddy, A.A. (1958). Composite nature of the flocculation process of top and bottom strains of *Saccharomyces*. *Journal of the Institute of Brewing* 64, 143.
Eddy, A.A. and Rudin, A.D. (1958a). Part of the yeast surface apparently involved in flocculation. *Journal of the Institute of Brewing* 64, 19-21.
Eddy, A.A. and Rudin, A.D. (1958b). Comparison of the respective electrophoretic and flocculation characteristics of different strains of *Saccharomyces*. *Journal of the Institute of Brewing* 64, 139-142.
Edwards, V.H. (1969). The recovery and purification of biochemicals. *Advances in Applied Microbiology* 11, 159-210.
Ellwood, D.C. and Tempest, D.W. (1972). Effects of environment on bacterial wall content and composition. *Advances in Microbial Physiology* 7, 83-117.
Finstein, M.S. (1967). Growth and flocculation in a Zoogloea culture. *Applied Microbiology* 15, 962-963.
Forster, C.F. (1968). The surface of activated sludge particles in relation to their settling characteristics. *Water Research* 2, 767-776.
Forster, C.F. and Choudhry, N.M. (1972). Physico-chemical studies in activated sludge bioflocculation. *Effluent and Water Treatment Journal* 12, 127-131.
Forster, C.F. and Lewin, D.C. (1972). Polymer interactions at activated sludge surfaces. *Effluent and Water Treatment Journal* 12, 520-525.
Friedman, B.A. and Dugan, P.R. (1968a). Concentration and accumulation of metal ions by the bacterium *Zoogloea*. *Developments in Industrial Microbiology* 9, 381-388.
Friedman, B.A. and Dugan, P.R. (1968b). Identification of *Zoogloea* species and relationship to zoogloeal matrix and floc

formation. *Journal of Bacteriology* 95, 1903-1909.
Friedman, B.A., Dugan, P.R., Pfister, R.M. and Remsen, C.C. (1968). Fine structure and composition of zoogloeal matrix surrounding *Zoogloea ramigera*. *Journal of Bacteriology* 96, 2144-2153.
Friedman, B.A., Dugan, P.R., Pfister, R.M. and Remsen, C.C. (1969). Structure of exocellular polymers and their relationship to bacterial flocculation. *Journal of Bacteriology* 98, 1328-1334.
Gasner, L.L. and Wang, D.I.C. (1970). Microbial cell recovery enhancement through flocculation. *Biotechnology and Bioengineering* 12, 873-887.
Geilenkotten, I. and Nyns, E.J. (1971). The biochemistry of yeast flocculence. *The Brewers Digest, April*, 64-70.
Gregory, J. (1973). Rates of flocculation of latex particles by cationic polymers. *Journal of Colloid and Interface Science* 42, 448-456.
Harrisberger, M. (1976). Immobilization of protein and polysaccharide on magnetic particles: Selective binding of microorganisms by Concanavalin A - magnetite. *Biotechnology and Bioengineering* 18, 1647-1651.
Harris, R.H. and Mitchell, R. (1973). The role of polymers in microbial aggregation. *Annual Review of Microbiology* 27, 27-50.
Harris, R.H. and Mitchell, R. (1975). Inhibition of the flocculation of bacteria by biopolymers. *Water Research* 9, 993-999.
Hattori, R., Hattori, T. and Furusaka, C. (1972). Growth of bacteria on the surface of an ion-exchange resin. *Journal of General and Applied Microbiology* 18, 271-284.
Hodge, H.M. and Metcalf, S.N. (1958). Flocculation of bacteria by hydrophilic colloids. *Journal of Bacteriology* 75, 258-264.
Israelachvili, J.N. (1973). Van der Waals forces in biological systems. *Quarterly Review of Biophysics* 6, 341-387.
Israelachvili, J.N. and Ninham, B.W. (1977). Intermolecular forces - the long and short of it. *Journal of Colloid and Interface Science* 58, 14-25.
Jack, T.R. and Zajic, J.E. (1977). The immobilization of whole cells. *Advances in Biochemical Engineering* 5, 125-137.
Janssen, P.E., Kenne, L. and Lindberg, B. (1975). Structure of the extracellular polysaccharide from *Xanthamonas campestris*. *Carbohydrate Research* 45, 275-282.
Jeanes, A. (1974). Application of extracellular microbial polysaccharide - polyelectrolytes. Review of literature, including patents. *Journal of Polymer Science, Polymer Series* 45, 209-227.
Jensen, H.E. and Mendlik, F. (1951). A study of yeast flocc-

ulation. *Proceedings, European Brewing Convention, Brighton* pp. 59-81.

Jones, G.E. and Starkey, R.L. (1961). Surface-active substances produced by *Thiobacillus thiooxidans*. *Journal of Bacteriology* 82, 788-789.

Kelly, D.P. (1976). Extraction of metals from ores by bacterial leaching: Present status and future prospects. In *Microbial Energy Conversion*, pp. 329-338. Edited by M. Schlegel and J. Barnea. Gottingen: Erich Goltze KG.

Kennedy, J.F., Barker, S.A. and Humphreys, J.D. (1976). Microbial cells living immobilized on metal hydroxides. *Nature, London* 261, 242-244.

Kiuchi, K., Kuraishi, H., Murooka, H., Aida, K. and Uemura, T. (1968). Floc formation in activated sludge. *Journal of General and Applied Microbiology* 14, 387-397.

Lee, C.K. and Long, R.E. (1974). Enzymatic process using immobilized microbial cells. United States of America Patent Number 3 821 086.

Lindquist, W. (1953). On the mechanism of yeast flocculation. *Journal of the Institute of Brewing* 59, 59-61.

Lips, A. (and Jessup, N.E.) (1978). This volume pp. 5-27.

Long, J.A., Osmond, D.W.J. and Vincent, B. (1972). The equilibrium aspects of weak flocculation. *Journal of Colloid and Interface Science* 42, 545-553.

Lyklema, J. and Norde, W. (1973). Biopolymer adsorption with special reference to the serum albumin-polystyrene latex system. *Croatica Chemica Acta* 45, 67-83.

Lyons, T.P. and Hough, J.S. (1971). Further evidence for cross-bridging hypothesis for flocculation of brewer's yeast. *Journal of the Institute of Brewing* 77, 300-305.

Martin, C.K.A. and Perlman, D. (1976). Conversion of L-Sorbose to L-Sorbosone by immobilized cells of *Gluconobacter melanogenus* IFO 3293. *Biotechnology and Bioengineering* 18, 217-237.

Matile, Ph., Morr, H. and Robinow, C.F. (1969). Yeast cytology. In *The Yeasts*, Volume I, pp. 220-302. Edited by A. H. Rose and J.A. Harrison, London : Academic Press.

McGoran, C.J.M., Duncan, D.W. and Walden, C.C. (1969). Growth of *Thiobacillus ferrooxidans* on various substrates. *Canadian Journal of Microbiology* 15, 135-138.

McGregor, I.C. and Finn, R.K. (1969). Factors affecting the flocculation of bacteria by chemical additives. *Biotechnology and Bioengineering* 11, 127-128.

McKinney, R.E. (1952). A fundamental approach to the activated sludge process. II. A proposed theory of floc formation. *Sewage and Industrial Wastes* 24, 280-287.

McKinney, R.E. (1956). In *Biological treatment of Sewage and Industrial Wastes*, Volume 1, pp. 88-100. Edited by J. Mc-

Cabe and W.W. Eckenfelder. New York : Reinhold.
McKinney, R.E. and Horwood, M.P. (1952). Fundamental approach to the activated sludge process. I. Floc producing bacteria. *Sewage and Industrial Wastes* 24, 117-123.
McKinney, R.E. and Weichlein, R.G. (1953). Isolation of floc-producing bacteria from activated sludge. *Applied Microbiology* 1, 259.
McLaren, A.D. (1954). The absorption and reaction of enzymes and proteins on kaolinite. *Journal of Physical Chemistry* 58, 129-137.
Metz, B. and Kossen, N.W.F. (1977). The growth of molds in the form of pellets - a literature review. *Biotechnology and Bioengineering* 19, 781-799.
Mill, P.J. (1964). The effect of nitrogenous substances on the time of flocculation of *Saccharomyces cerevisiae*. *Journal of General Microbiology* 35, 53-68.
Mill, P.J. (1966). Phosphomannans and other components of flocculent and non-flocculent walls of *Saccharomyces cerevisiae*. *Journal of General Microbiology* 44, 329-341.
Morris, E.O. (1966). Aggregation of unicells: Yeasts. In *The Fungi*, Volume 2, pp. 63-82. Edited by G.C. Ainsworth and A.S. Sussman. New York : Academic Press.
Mueller, J.A., Morand, J. and Boyle, W.C. (1967). Floc sizing techniques. *Applied Microbiology* 15, 125-134.
Munson, R.J. and Bridges, B.A. (1964). Take-over. An unusual selection process in steady state cultures of *Escherichia coli*. *Journal of General Microbiology* 37, 411-418.
Murr, L.E. and Berry, V.K. (1976). An electron microscope study of bacterial attachment to chalcopyrite: Micro-structural aspects of leaching. In the *Extractive Metallurgy of Copper*, pp. 670-689. Edited by J.C. Yannopolous and J.C. Aggarwal, New York : American Institute of Mining Engineers.
Napper, D.H. (1977). Steric stabilization. *Journal of Colloids and Interface Science* 58, 390-407.
Napper, D.H. and Hunter, R.J. (1972). Hydrosols, MTI International Review of Science, Physical Chemistry, Series I, Volume 7, *Surface Chemistry and Colloids*, pp. 241-306. Edited by M. Kerker, London : Butterworth.
Navarro, J.M. and Durand, G. (1977). Modification of yeast metabolism by immobilization onto porous glass. *European Journal of Applied Microbiology* 4, 243-254.
Northrop, J.H. and DeKruif, P.H. (1921). The stability of bacterial suspensions. *Journal of General Physiology* 4, 639-654.
Parker, D.S., Kaufmann, W.J. and Jenkins, D. (1971). Physical conditioning of the activated sludge floc. *Journal Water Pollution Control Federation* 43, 1817.

Parsegian, V.A. (1973). Long-range physical forces in the biological milieu. *Annual Review of Biophysics and Bioengineering* 2, 221-225.

Pavoni, J.L., Tenney, M.W. and Echelberger, W.F. (1972). Bacterial exocellular polymers and biological flocculation. *Journal Water Pollution Control Federation* 44, 414-431.

Peter, G. and Wuhrmann, K. (1970). Contribution to the problem of bioflocculation in the activated sludge process. *Proceedings of the 5th International Water Pollution Research Conference*. Oxford : Pergamon Press.

Pethica, B.A. (1961). The physical chemistry of cell adhesion. *Experimental Cell Research*, supplement 8, 123-140.

Rainbow, C. (1966). Flocculation of brewer's yeast. *Process Biochemistry* 1, 489-492.

Rainbow, C. (1970). Brewer's Yeast. In *The Yeasts*, Vol. 3, pp. 147-224. Edited by A.H. Rose and J.S. Harrison. London : Academic Press.

Rees, D.A. and Welsh, E.J. (1977). Secondary and tertiary structure of polysaccharides in solution and gels. *Angewandte Chemie International Edition in English* 16, 214-224.

Roberts, K., Wennerberg, A.-M. and Friberg, S. (1974). The influence of added saccharide, protein and lipid on the sedimentation of *E. coli* bacteria using aluminium sulphate and polyacrylamide. *Water Research* 8, 61-65.

Sarkar, N. and Teot, A.S. (1973). Coagulation of negatively-charged colloids, by anionic polyelectrolytes and metal ions. *Journal of Colloid and Interface Science* 43, 370.

Schaeffer, W.I., Holbert, P.E. and Umbreit, W.W. (1962). Attachment of *Thiobacillus thiooxidans* to sulphur crystals. *Journal of Bacteriology* 85, 137-140.

Schaeffer, W.I. and Umbreit, W.W. (1963). Phosphotidylinositol as a wetting agent in sulphur oxidation by *Thiobacillus thiooxidans*. *Journal of Bacteriology* 85, 492-493.

Stanley, S.O. and Rose, A.H. (1967). On the clumping of *Corynebacterium xerosis* as affected by temperature. *Journal of General Microbiology* 48, 9-23.

Starkey, R.L., Jones, G.E. and Frederick, L.R. (1956). Effects of medium agitation and wetting agents on the oxidation of sulphur by *Thiobacillus thiooxidans*. *Journal of General Microbiology* 15, 329-334.

Sutherland, I. (1975). The bacterial wall and surface. *Process Biochemistry*, April, 4-8.

Taylor, N.W. and Orton, W.L. (1973). Effect of alkaline earth metal salts on flocculence in *Saccharomyces cerevisiae*. *Journal of the Institute of Brewing* 79, 294-297.

Temple, K.L. and Koehler, W.A. (1954). Drainage from bituminous coal mines. *West Virginia University Bulletin*, Series SY 4, 1.

Tenney, M.W. and Stummn, W.J. (1965). Chemical flocculation of microorganisms in biological water treatment. *Journal Water Pollution Control Federation* 32, 1370-1388.

Tenney, M.W. and Verhoff, F.H. (1973). Chemical and autoflocculation of microorganisms in biological waste water treatment. *Biotechnology and Bioengineering* 15, 1045-1073.

Tezuka, Y. (1967). Magnesium ion as a factor governing bacterial flocculation. *Applied Microbiology* 15, 1256.

Tezuka, Y. (1969). Cation-dependent flocculation in a *Flavobacterium* species predominant in activated sludge. *Applied Microbiology* 17, 222-228.

Topiwala, H.H. and Hamer, C. (1971). Effect of wall growth in steady state continuous cultures. *Biotechnology and Bioengineering* 13, 919-922.

Torma, A.E. (1977). The role of *Thiobacillus ferrooxidans* in hydrometallurgical processes. *Advances in Biochemical Engineering* 6, 1-38.

Treweek, G.P. and Morgan, J.J. (1977). Polymer flocculation of bacteria. The mechanism of *E. coli* aggregation by polyethyleneimine. *Journal of Colloid and Interface Science* 40, 258-273.

Unz, R.F. and Dondero, N.C. (1967). The Predominant Bacteria in Natural Zoogloeal Colonies. I. Isolation and Identification. *Canadian Journal of Microbiology* 13, 1671-1682.

Verwey, E.J.W. and Overbeek, J. Th. G. (1948). *Theory of the Stability of Lyophobic Colloids*. Amsterdam : Elsevier.

Vincent, B. (1974). The effect of adsorbed polymers on dispersion stability. *Advances in Colloid and Interface Science* 4, 193-277.

Vogler, K.G. and Umbreit, W.W. (1941). The necessity for direct contact in sulphur oxidation by *Thiobacillus thiooxidans*. *Soil Science* 51, 331-337.

Warren, G.H. and Gray, J. (1955). Studies on the properties of a polysaccharide constituent produced by *Pseudomonas aeruginosa*. *Journal of Bacteriology* 70, 152-157.

Weiss, R.L. (1973). Attachment of bacteria to sulphur in extreme environments. *Journal of General Microbiology* 77, 501-507.

Wessman, G.E. and Miller, D.J. (1966). Biochemical and physical changes in shaken suspensions of *Pasteurella pestis*. *Applied Microbiology* 14, 636-642.

Whitaker, A. and Long, P.A. (1973). Fungal pelleting. *Process Biochemistry* 7 (11), 27-31.

Wiersema, P.H., Loeb, A.L. and Overbeek, J. Th. G. (1966). Calculation of the electrophoretic mobility of a spherical colloid particle. *Journal of Colloid and Interface Science* 22, 78-99.

Wilkinson, T.G. and Hamer, G. (1974). Wall growth in mixed

bacterial cultures growing on methane. *Biotechnology and Bioengineering* 16, 251-260.

THE ATTACHMENT OF BACTERIA TO SURFACES IN AQUATIC ENVIRONMENTS

MADILYN FLETCHER

Department of Environmental Sciences,
University of Warwick, Coventry CF4 7AL.

INTRODUCTION

A large proportion of aquatic bacteria are found attached to submerged solid surfaces, such as particulate detritus, man-made structures and other micro- or macro-organisms. Attached bacterial populations are particularly significant in low-nutrient waters, where numbers of suspended organisms are generally quite low.

The formation of attached bacterial 'films' on submerged surfaces has some very important economic and ecological implications. Firstly, primary film formation by bacteria may affect the subsequent attachment, and thus often development, of larger organisms, such as macroalgal spores or invertebrate larvae. In this way, bacterial attachment can indirectly influence the costly fouling and corrosion of man-made structures [ZoBell and Allen, 1935; Corpe, 1970a] or the production of commercial food crops, such as oysters [Mitchell and Young, 1972]. Secondly, attached bacteria are important in the biodegradation of organic substances and are obviously so in the breakdown of particulate material. Finally, attached bacteria play an important, albeit poorly understood, role in the maintenance of the stability of the overall ecosystem. This includes the relationship between epiphytic and epizoic bacteria and their hosts, and the ability of attached communities to act as microbial reservoirs, able to repopulate surrounding waters when environmental conditions are favourable for growth.

In most environments there are attached and free-living bacterial populations, so that neither community is entirely predominant. Whether an individual bacterium becomes attached is dependent upon both its attachment mechanism and the environmental conditions. A number of modes of attachment have

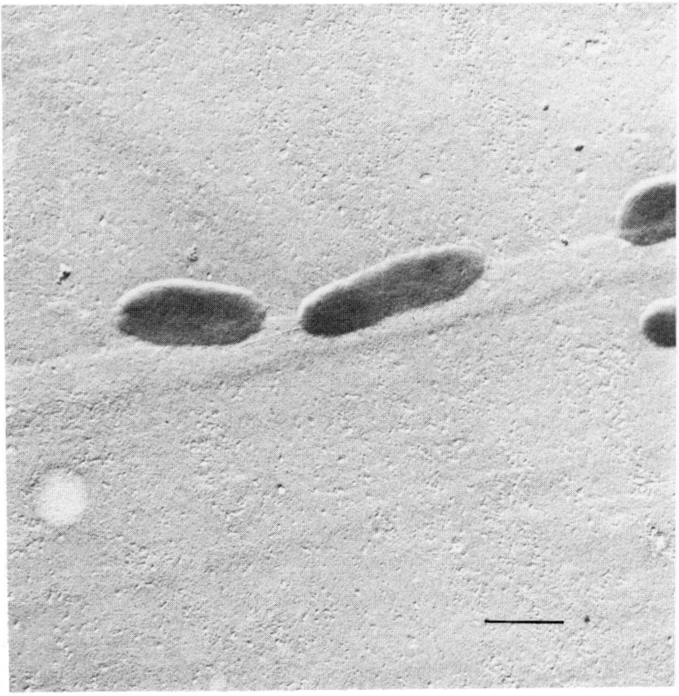

Fig. 1. Bacteria from sea water which have been attached to a Formvar-coated electron microscope grid. Fixed in formalin vapour and shadowed with gold- palladium. Bar marker represents 1 μm.

been described [Hirsch and Pankratz, 1970]; these include special structures, such as pili which are found on some members of the Enterobacteriaceae and Pseudomonadaceae, holdfasts which occur on some prosthecate bacteria such as caulobacters, and the deposition of inorganic cements.

However, most aquatic bacteria have no obvious means of attachment and simply seem to 'stick' to the substratum (Figure 1). These bacteria are apparently attached by means of extracellular adhesives, which may exist as thin cell-surface coats or as a diffuse intercellular matrix, retaining colonies of cells on the surface [Fletcher and Floodgate, 1973]. Evidence from histochemical electron microscopy has suggested that

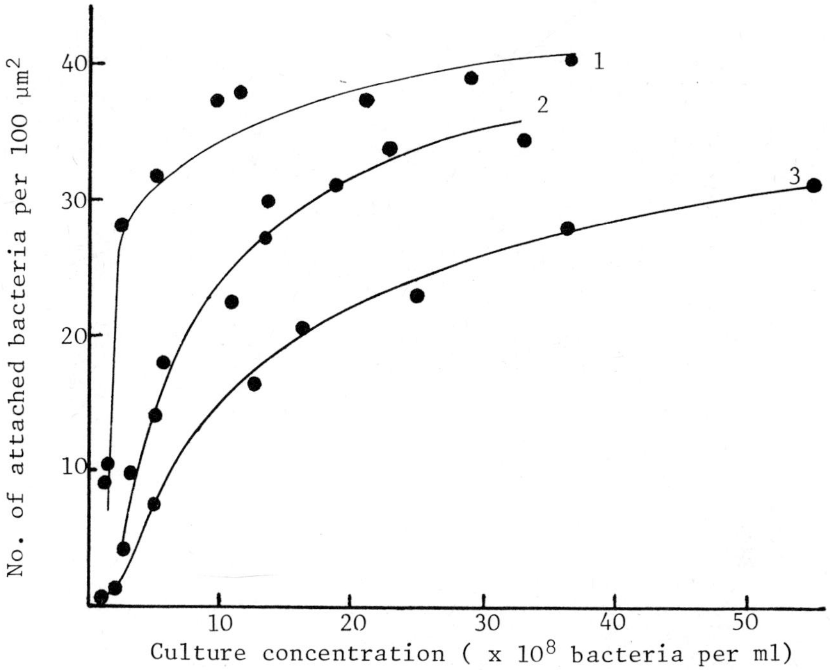

Fig. 2. The relationship between the concentration of suspended cells and the number of cells which become attached to polystyrene after 2 hours. 1 - log phase, 2 - stationary phase, 3 - death phase. (From Fletcher, 1977).

the adhesive is an acidic polymer [Fletcher and Floodgate, 1973]. Further acidic polysaccharides [Corpe, 1970b] and acidic glycoproteins [Corpe, Matsuuchi and Armbruster, 1976] have been isolated from the supernatants of cultures of adhesive marine bacterial strains.

BACTERIAL ATTACHMENT BY MEANS OF EXTRACELLULAR POLYMERS

If it is assumed that bacterial attachment is mainly mediated by the extracellular polymeric material two conditions must be met. The bacterium must encounter and come close enough to a solid surface for the extracellular polymer to bridge the separating space and, secondly, the extracellular polymeric adhesive must be adsorbed onto the surface.

The probability that a bacterium will encounter a submerged surface is dependent upon a number of factors. These include both the concentration of suspended cells and the time which the suspension is exposed to the surface. Figure 2 shows the relationship between the number of attached bacteria

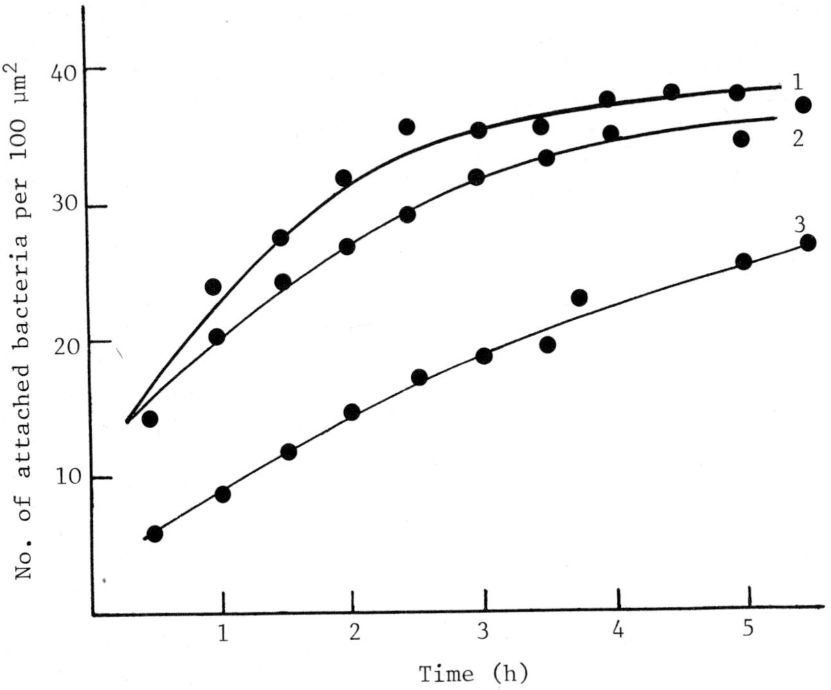

Fig. 3. The relationship between the time allowed for attachment and the number of attached cells on polystyrene. 1 = *log phase,* 2 = *stationary phase,* 3 = *death phase. (From Fletcher, 1977).*

and the suspended cell concentration; there is a positive correlation until the curve begins to level off as the surface (polystyrene) becomes covered with bacteria. The organism used in this experiment, as well as in other studies specifically discussed below, was a marine *Pseudomonas* sp. (NCMB 2021).

Figure 3 shows results from a similar experiment which demonstrated a relationship between the number of attached bacteria and the time allowed for attachment. An increase in either culture concentration or time has led to an increase in opportunity for attachment by increasing the probable number of bacterial collisions with the surface. Another factor affecting the chance of encountering a surface is cell motility. In the experiments mentioned above, there was a progressive decrease in motility after the onset of the stationary phase, and this could account for the decrease in attached numbers which corresponded with increasing culture age (Figures 2 and 3). Obviously, motile cells are more likely to meet a surface than are non-motile cells, which are dependent upon Brownian motion or water currents. (The importance of motility is further dealt with later).

It is not enough, however, that a bacterium approaches a surface. It must also be near enough for the extracellular polymer to absorb and this could be prevented by electrostatic repulsion forces existing between the bacterial surface and the potential substratum. Such forces exist between surfaces of the same net charge, and since most bacteria [Harden and Harris, 1953] and most surfaces carry a net negative surface charge, repulsion forces must be considered important.

This was recently demonstrated [Heckels, Blackett, Everson and Ward, 1976] when the attachment properties of *Neisseria gonorrhoeae* were altered after the cell surface charge was modified by chemical treatment. The force of repulsion is directly related to the surface potentials and the radii of curvature of the two surfaces, and decreases as the distance between the two surfaces is increased.

The influence of electrostatic repulsion on bacterial attachment has been indirectly indicated by the influence of cation concentration on attachment. When a charged surface is immersed in an aqueous solution, counter-ions are attracted to the surface, so that the so-called diffuse electrical double layer is formed [Shaw, 1970]. The thickness of the double layer is decreased through compression as the solution electrolyte concentration or valency is increased, and repulsion energies are thereby reduced. Accordingly, Marshall, Stout and Mitchell [1971] observed an increase in the attachment of selected marine bacteria to glass as the electrolyte concentration in the medium was increased.

However, it is often difficult to interpret data from studies which experimentally alter cation concentrations since certain cations (for example, Na^+, Ca^{2+}, Mg^{2+}) enter into specific physiological activities so that the bacterial population will remain stable only within a certain range of cation concentrations; this is particularly marked with marine bacteria [MacLeod, 1965] and halophiles. Secondly cations may interact with acidic polymers and thus possibly denature the adhesive.

To test the effect of electrolyte concentration on attachment of a marine pseudomonad (NCMB 2021), Al^{3+} and La^{3+} were added (as $AlCl_3$ and $LaCl_3$), at concentrations ranging between 17 and 200 μM, to suspensions of attaching cells. With both cations, there was a progressive decrease in attachment with increase in cation concentration, and La^{3+} was the more

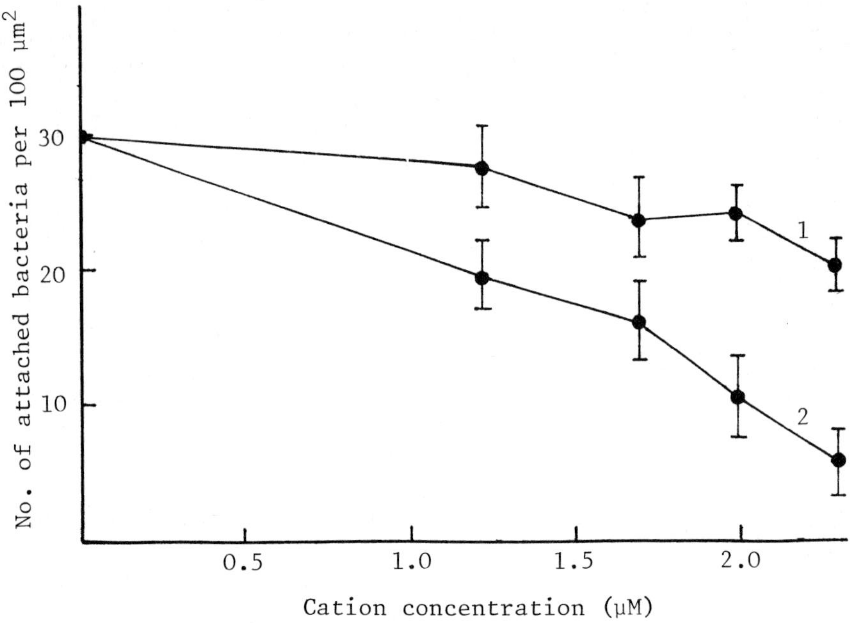

Fig. 4. The effect of aluminium ions and lanthanum ions on bacterial attachment to polystyrene. 1 = *aluminium ions,* 2 = *lanthanum ions.*

effective of the two in reducing attachment (Figure 4). The bacteria seemed otherwise unaffected by the cations. This result was somewhat surprising since an increase in concentration of trivalent cations would result in compression of the electrical double layer, and thus reduction in repulsion forces and increase in attachment. Since attachment was reduced, it suggests that Al^{3+} and La^{3+} interacted with the acidic adhesive polymer [Fletcher and Floodgate, 1973], altering its adhesive properties. Previous studies [Fletcher and Floodgate, 1976] have suggested that Ca^{2+} and Mg^{2+} may also interact with polymers produced by the organism.

Cell motility may also be important, in that a motile bacterium may have sufficient kinetic energy to overcome electrostatic repulsion forces. The effect of motility on attachment has been explored by investigating the relationship between the proportion of motile cells and the resulting number of attached cells. A marine pseudomonad (NCMB 2021) was grown in sea water, supplemented with peptone and yeast extract (0.01% of each) [Fletcher, 1976] for 17 h, at which time the organisms were in the log phase of growth, and there was a high proportion of motile organisms. The bacteria were collected by cen-

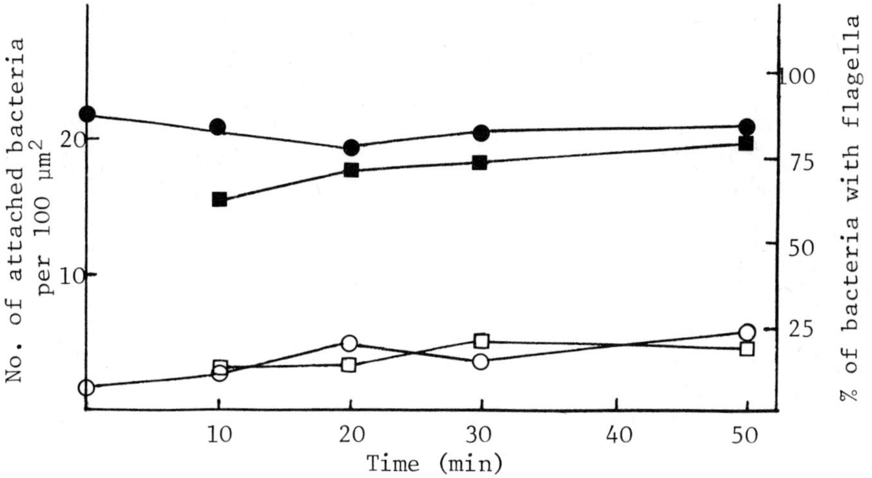

Fig. 5. The attachment of motile and non-motile bacteria to polystyrene. Percentage of bacteria with flagella : ○ = homogenised culture, ● = control. Number of attached bacteria : □ = homogenised culture, ■ = control.

trifugation, resuspended in autoclaved, filtered (0.2 μm porosity) sea water, and part of the resultant suspension was treated to remove the flagella from the cells (2.5 min at approximately 2/3 full speed in an MSE homogeniser). At this time, no motile cells could be detected microscopically. The homogenising probably had no adverse effect other than removal of flagella, since there was a temperature rise of only 1° C during the process; furthermore bacteria quickly regained motility (occasionally within 20 min) as new flagella were produced.

These homogenised non-motile cells were then added, in 30 ml portions, to polystyrene Petri dishes (Sterilin), which served as the attachment surface. A series of dishes containing untreated motile cells was also prepared as a control. At intervals of 10, 20, 30 and 50 min, dishes were rinsed with sterile sea water, and the bacteria which were attached to the dishes were fixed with Bouin's fixative, stained with ammonium oxalate crystal violet and counted microscopically. At the same time, a sample of bacterial suspension was removed from each dish, and a flagella stain [Clark, 1976] was carried out on each sample to estimate the proportion of potentially motile cells.

The data, shown in Figure 5, demonstrate that cultures of motile cells were able to attach in higher numbers than those in which flagella had been removed from the bacteria. Motil-

Table 1.

Effect of stirring on bacterial attachment

Time allowed for attachment (min)	Number of attached bacteria per 100 μm^2 ± sd. Stirred	Control
3	0.66 ± 0.5	0.99 ± 0.5
5	1.04 ± 0.7	1.10 ± 0.4
8	0.93 ± 0.6	1.0 ± 0.4
12	0.80 ± 0.4	1.3 ± 0.4
15	0.9 ± 0.4	1.4 ± 0.4

Pseudomonads were cultured for 17 h, and flagella were removed by homogenisation (see text). Treated cells were adjusted to concentrations of 8.9 x 10^8 bacteria ml^{-1}, and 250 ml portions were added to 380 ml staining troughs. Polystyrene pieces (5mm x 75mm x 25mm) cut from Petri dishes and held in a glass rack, were provided for attachment surfaces. The suspension was stirred with a magnetic stirrer at as fast a rate as possible, without creating a vortex or causing displacement of the slides. Slides were removed at intervals, washed with sterile sea water, fixed, stained and the attached bacteria were counted microscopically. An unstirred control was provided.

ity does not seem to increase attachment simply by increasing the number of bacterial encounters with the substratum, since the attachment of non-motile cells was not increased by stirring the cell suspension (Table 1). However, if the force with which a non-motile cell encounters a surface is artificially increased by centrifugation, the numbers of bacteria which attach are directly related to the force applied if the centrifugal force is at 90° to the polystyrene surface.

This was demonstrated by removing the flagella from the bacteria (as above) and centrifuging the cells down onto polystyrene surfaces; the suspension was then held for 1 min at the appropriate relative centrifugal force. The results are shown in Figure 6. Centrifugation may facilitate the attachment of non-motile cells by forcing the cells closer towards the attachment surface, thus overcoming electrostatic repulsion forces, or by increasing the number of cells encountering the surface, since at higher values of g, more cells will be brought down from suspension to the attachment surface.

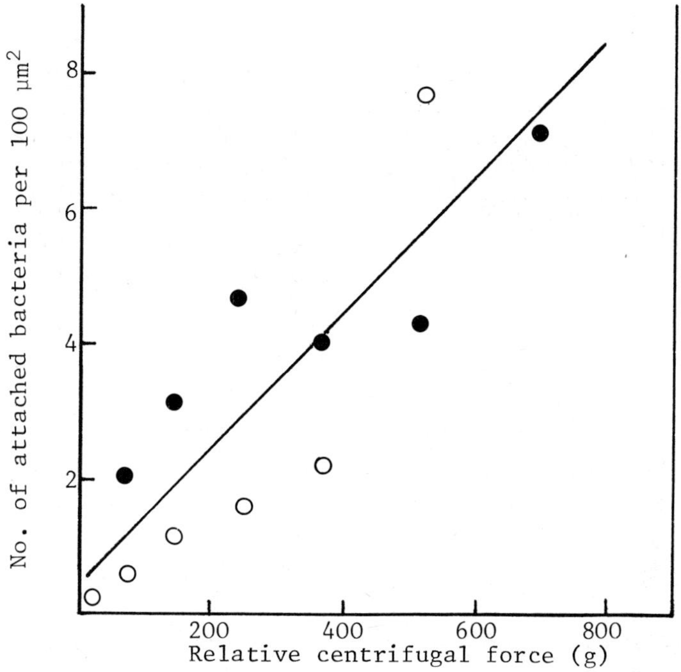

Fig. 6. The relationship between the relative centrifugal force applied normal to the attachment surface and the number of non-motile bacteria which became attached. Different symbols represent separate experiments.

If the bacteria are not able to overcome electrostatic repulsion barriers, attachment may still be possible, as is described by the DLVO theory of lyophobic colloid stability [Derjaguin and Landau, 1941; Verwey and Overbeek, 1948; Shaw, 1970]. In this model, which has been used to explain the adhesion of negatively charged tissue cells to a negative substrate [Curtis, 1967], the cell is held at a finite distance 1 - 10 nm [Maroudas, 1975] from the substratum, where, under the right conditions, London-van der Waals-type attraction forces may be able to predominate over electrostatic repulsion forces.

Marshall *et al.* [1971] used the DLVO theory to account for the 'reversible sorption' of bacteria to surfaces. They suggested that this reversible sorption is an initial stage of bacterial attachment, during which the cells seem to 'settle' on a substratum, but still exhibit Browian motion and are easily dislodged by washing. Many bacteria 'swim away' spontaneously, but some of these cells eventually become permanently attached, and this time-dependent stage of 'irreversible sorption' is probably due to the production of extracellular poly-

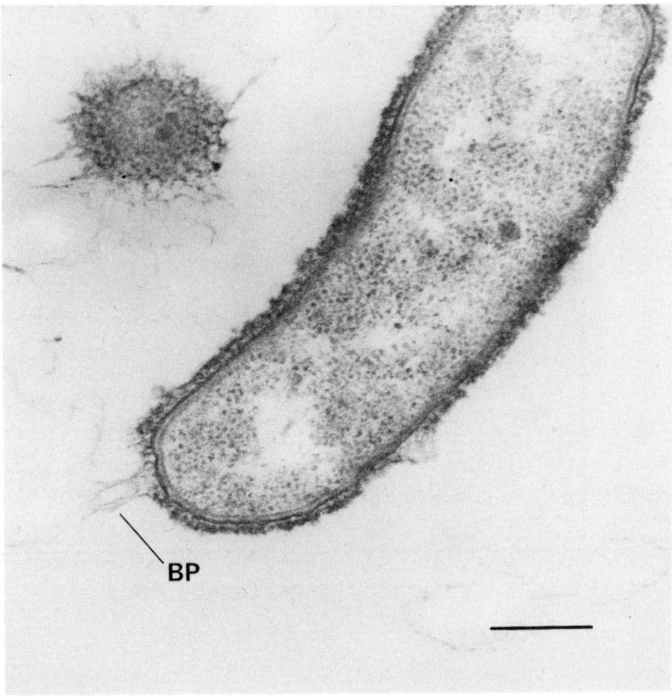

Fig. 7. A marine pseudomonad attached to a Millipore filter. The gap between bacterial and filter surface is bridged by the cell surface polymer (BP). Fixed with glutaraldehyde/osmium/ruthenium red; post-stained with lead citrate and uranyl acetate. Bar marker represents 0.25 μm.

mers which can bridge the gap between bacterial and substratum surfaces [Marshall *et al.*, 1971].

It is unlikely that the DLVO theory can fully explain bacterial attachment to surfaces, since it considers only electrostatic and dipole-type interactions and excludes more specific interactions, such as hydrogen and hydrophobic bonding [Maroudas, 1975]. The deficiencies of the DLVO model are further emphasised by experimental evidence, such as electron micrographs demonstrating polymers bridging the bacterial and substratum surfaces (Figure 7), and demonstration of the importance of non-electrostatic free energy effects (see below).

At the surfaces of all liquids and solids there is a surf-

Table 2.

Surface and interfacial data for selected surfaces (from Andrade, 1973)

Surface	γ_c	γ_{SW}	W_{SW}
Glass	72.8	23.8	219.0
Water	-	0	145.6
Polyethylene terephthalate	43	31.6	70.6
Polystyrene	33	34.5	78.9
Polyethylene	31	38.0	67.2
Air	0	72.8	0

The critical surface tension (γ_c) is a measure of the surface free energy, whereas the interfacial free energy (γ_{SW}) is a measure of the residual bonding potential at the surface/water interface. The work of adhesion (W_{SW}) indicates the interfacial attraction or bonding between the surface and water. The units are erg cm^{-2}.

ace free energy, which is a measure of the unsatisfied bonding capacity of the surface [Andrade, 1973]. There are a number of different types of force which can contribute to surface bonding potential, including dispersion, dipole, acid-base and metallic type forces [Andrade, 1973]. There is always the tendency for surfaces to obtain a minimum free energy by satisfying the potential bonding capacity of the surface and one way this can be done is through the adsorption of substances (for example, bacterial surface polymers) onto the surface. Thus adsorption of components onto surfaces is favoured by a reduction in the free energy of the system and is determined by whether the two molecular phases involved possess common forces (for example, dispersion, dipole) and thus can interact specifically.

Surface free energies are good approximations for solid/gas or liquid/gas interfaces, but when a surface is submerged in a liquid, the potential bonding capacities of the solid and liquid surfaces may be partially satisfied. It is then the residual bonding capacity, or the interfacial energy, which becomes important in determining adsorption of dissolved medium components. Generally (see Table 2), there is a progressive increase in the interfacial tension between water and polymer surfaces with critical surface tensions (a measure of surface free energy; Zisman, 1964) less than 72.8 erg cm^{-2} (the surface tension of water). Another important factor to consider is the attraction, or work of adhesion (Table 2), between the surface and water, since water strongly adsorbed to a surface may present a barrier to the subsequent adsorption of other components.

The importance of surface free energies, hence interfacial energies, on bacterial attachment was shown in an extension of a study [Fletcher and Loeb, 1976] of the attachment of a marine pseudomonad (NCMB 2021) to a variety of surfaces. The results (Figure 8) are presented in histograms showing the frequency distribution of numbers of attached bacteria on the different substrate; where known, the critical surface tension (γ_c), which is a measure of surface free energy, is also given.

From the data, three main points are evident, firstly, the bacteria attached in high numbers to the low energy surfaces (fluoropolymer, polyethylene, polystyrene, polyethylene terephthalate), which also have a very low negative surface charge; secondly, very few bacteria attached to the higher energy surfaces which also have an appreciable negative charge (processed plastics, glass, mica), and lastly, attachment to high energy surfaces with positive (platinum) or very low negative (germanium) surface charge was moderate, and at a level between those of the previous two. It is significant that the highest numbers of attached bacteria were obtained with low energy plastics, since the interfacial energies between these surfaces and water would be higher than those between glass and water (Table 2). Bacteria also tend to accumulate more extensively at liquid/liquid interfaces of high interfacial energies than those with lower energies [Marshall, 1976].

In a further experiment, the bacteria were allowed to attach to polyethylene which was prepared by recrystallising it against gold foil [according to Schonhorn, 1968]. This provided many nucleation sites so that the crystallinity and density of the plastic was increased at the surface. Since free energy is increased with both density and crystallinity, the gold foil-recrystallised polyethylene has a higher free energy than ordinary polyethylene and it was assumed that this did not result in an increase in surface charge. The results are shown in Figure 9, which includes comparative data on attachment to polystyrene and to a high purity epoxy resin; there is a decrease in the number of attached bacteria with the increase in substratum free energy (indicated by γ_c).

Experimental results indicate than that the attachment of bacteria is affected by two important substratum characteristics. Firstly, the surface charge of the substratum influences attachment through electrostatic interactions, presumably by

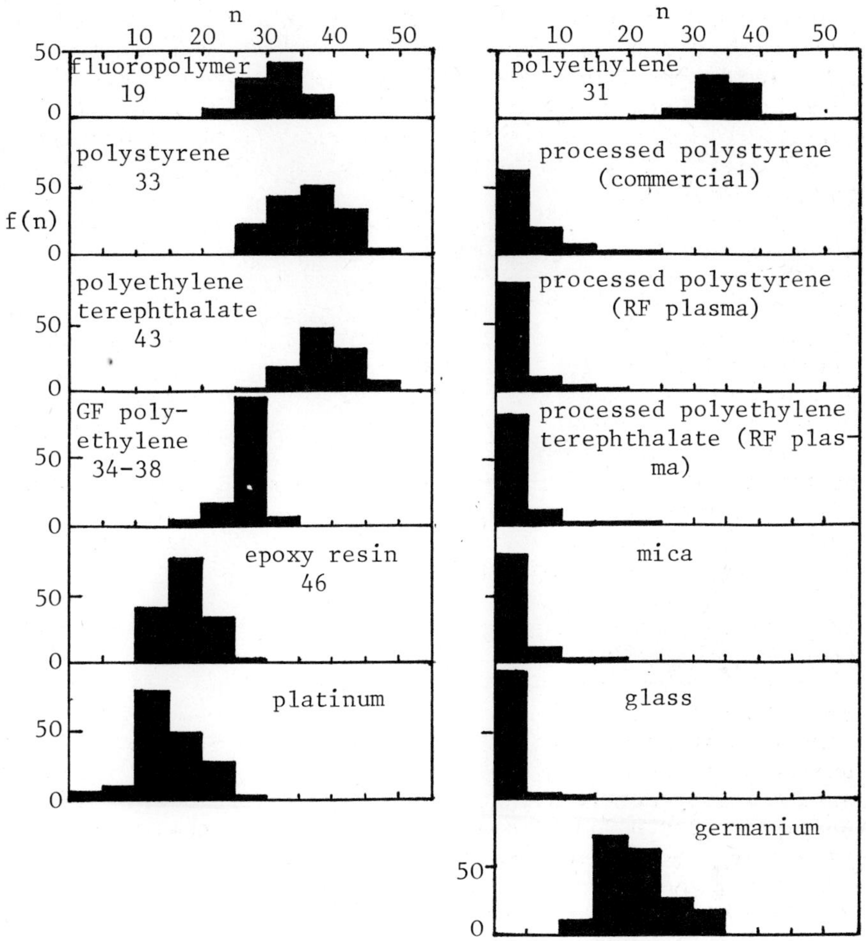

Fig. 8. Histograms representing the distribution of bacteria which attached to various substrata after 2 h. n = *no. of bacteria (at intervals of 5) per 100 μm² field;* f(n) = *no. of fields counted containing* n *bacteria. GF polyethylene is polyethylene recrystallised against gold foil (see text). Processed polystyrene (commercial) refers to tissue culture dishes, whereas processed plastics (RF plasma) have been treated in a radio frequency plasma cleaning device to increase surface charge density and free energy (Fletcher and Loeb, 1976). The figures in some of the histogram blocks are, where known, the critical surface tensions in erg cm^{-2}, and indicate the relative surface free energies of the materials.*

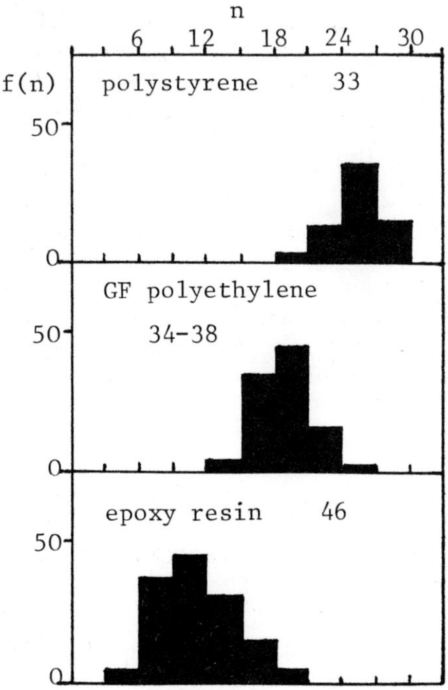

Fig. 9. Histograms representing the distribution of bacteria which attached to various substrata after 15 min. n = *number of bacteria (at intervals of 3) per 100* μm^2 *field;* f(n) = *number of fields counted containing* n *bacteria. The figures in the histogram blocks are the critical surface tensions in erg* cm^{-2}*, and indicate the relative surface free energies of the materials.*

inhibiting or facilitating the approach of the bacterium towards the substratum. Thus the attachment of this bacterium to negatively charged surfaces, such as glass and mica, was inhibited. Secondly, the interfacial energy at the surface/medium boundary and the bonding potential of the surface will affect the actual adsorption of the bacterial surface adhesive onto the substratum. This considers both the specific interactions between the adhesive and potential substratum, as well as the effect water may have in competing for bonding sites on the substratum.

Unfortunately, the situation is not always so simple, particularly in natural environments, where apparent substratum characteristics may be considerably altered by the adsorption of dissolved milieu components. The surface which the bacterium 'sees' may be quite different from the originally immersed material. It has been shown that, when adsorbed onto the attachment substratum, proteins, such as bovine serum albumin, gelatin, fibrinogen and pepsin, inhibit subsequent bacterial attachment [Fletcher, 1976]. A medium filtrate has in the same manner inhibited the attachment of this bacterium to platinum. The nature of an adsorbed molecular layer should depend to some extent on the characteristics of the substratum. Accordingly, Baier and Marshall found that the critical surface tension of siliconised germanium prisms which had been immersed in sea water differed from that of untreated germanium prisms under the same conditions; subsequent bacterial attachment to the two surfaces also differed [Marshall, 1976].

On the other hand, Loeb and Neihof [1975] found that when materials of markedly different surface charge were immersed in sea water, all acquired a moderately negative charge, apparently through the adsorption of macromolecular organic matter. Therefore, it is not at all easy to predict the composition of films which adsorb onto various surfaces, particularly in natural waters where the composition of dissolved components can be extremely complex and difficult to evaluate.

THE INFLUENCE OF SURFACES ON BACTERIAL ACTIVITY

One of the most interesting questions concerning bacterial attachment to surfaces deals with the effect of solid surfaces on bacterial activity. It is generally believed that bacterial activity is enhanced at surfaces, particularly in low nutrient environments, but this phenomenon is not at all understood, though the first experiments strongly suggesting a beneficial effect of solid surfaces were carried out on marine bacteria over forty years ago by ZoBell and Anderson [1936].

This work, along with subsequent studies by others, has been reviewed elsewhere [Corpe, 1970a; Atkinson and Fowler, 1974] and will not be detailed here; however, the main points are as follows:

(a) The addition of inert (not nutritive) solids can increase the activity of certain bacteria, for example, marine forms [ZoBell, 1943; Jannasch and Pritchard, 1972], *Escherichia coli* [Heukelekian and Heller, 1940], as indicated by an increase in suspended cell numbers or overall respiratory activity.

(b) The beneficial effect of solid surfaces usually occurs in very low-nutrient media [ZoBell, 1943; Jannasch and Pritchard, 1972], for example at less than 0.5 ppm glucose or peptone [Heukelekian and Heller, 1940].

(c) Actual attachment of cells is not always involved, and the association between bacteria and surfaces may be quite superficial [Heukelekian and Heller, 1940].

(d) The presence of solid surfaces does not always enhance bacterial activity [ZoBell, 1943] and, in some cases, the activity of free-living bacteria may exceed that of attached forms [Hattori and Furusaka, 1960].

However, if it is accepted that an increase in the surface area to volume ratio may increase the activity of certain aquatic bacteria, there are a number of possible explanations. First, nutrients (or growth factors) may be adsorbed and concentrated at surfaces, and thereby made more available to attached bacteria. This is the most commonly accepted explanation [Marshall, 1976], but although potential nutrients may be concentrated at interfaces, it is not certain whether a bacterium will be able to take up adsorbed substances. This is particularly true for macromolecular components, which tend not to enter into an adsorption equilibrium so that desorption may not occur [Kipling, 1965]. Secondly, more efficient use may be made of exoenzymes at surfaces. Possibly, adsorption of enzymes results in advantageous configurational changes or the diffusion of enzymes away from the surface may be slowed down [ZoBell, 1943].

Thirdly, the growth of suspended bacteria may be facilitated by the addition of solid surfaces if inhibitors are removed by such surfaces. This effect may be particularly noticeable in low-nutrient media with small bacterial populations, where there is little opportunity for the effects of inhibitors to be masked or compensated for by other 'dissolved' components, [Hardwood and Pirt, 1972]. Indeed, sea water has been found to be inhibitory to some marine bacteria, so that they are unable to survive if their numbers fall below a required population minimum [Jannasch, 1968].

Finally, a wide range of other chemical and physical factors will be different at the solid/liquid boundary, as compared with the bulk of the medium. These are primarily a function of the intermolecular forces between the two phases and of the interfacial energy, and include factors such as the hydrogen ion concentration and the redox potential.

Thus, it is extremely difficult to predict the effect of surfaces on bacterial activity, but some preliminary microautoradiographic studies have indicated that the activity of attached bacteria is affected by the composition of the underlying substratum [Fletcher, unpublished results].

Pseudomonads (for example, NCMB 2021) attached to polyethylene and to glass were incubated with a mixture of tritiated amino acids (10 μCi/ml) for 3 h. When the bacteria were attached to glass, there was considerable uptake of labelled amino acids by most of the cells; free-living cells from the same culture showed similar uptake (Figures 10 and 11). However with polyethylene the label did not seem to be particularly associated with the cells, and in many cases was randomly distributed over the surface (Figure 12).

A time series experiment and fixed-cell controls showed that the bacteria on polyethylene had in fact taken up the labelled amino acids, but that the label was being released from the cells by the end of the three hour incubation period. By

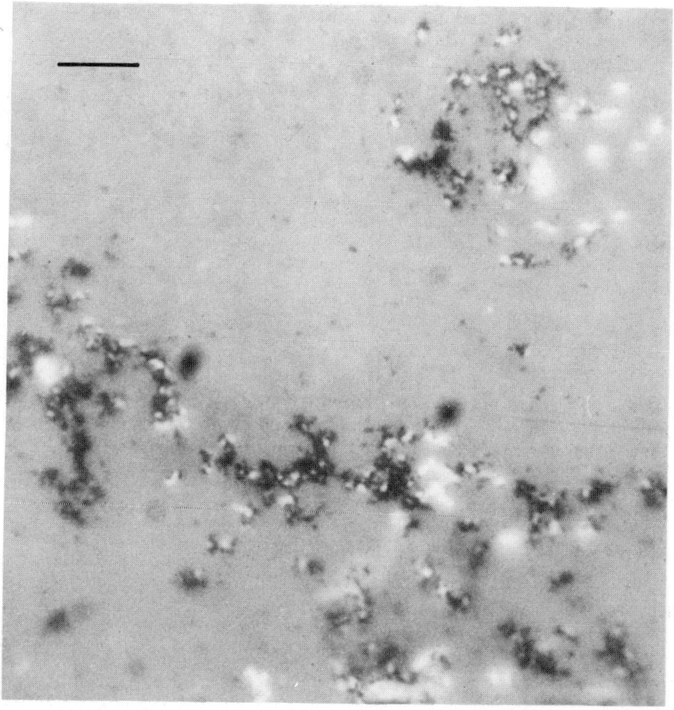

Fig. 10. Suspended cells which have been collected by filtration, prepared for autoradiography and stained with acridine orange. The bacteria appear white against the grey background. Most of the bacteria have taken up labelled amino acids, as indicated by the surrounding reduced silver grains (black) of the autoradiographic emulsion. Bar marker represents 5μm.

contrast, free-living bacteria and those on glass still retained much of the radioactive label. Labelled material randomly distributed on the polyethylene surface could then be adsorbed waste products or possibly recently synthesised extracellular polymer. Thus, bacteria on polyethylene appeared to be able to metabolise the amino acids at a faster rate than free-living bacteria or those attached to glass. Certainly the influence of surfaces on bacterial activity should be looked at much more closely in the future.

In natural aquatic environments, it is particularly difficult to predict the activity of attached bacteria, since they are usually found in larger communities with other micro- and

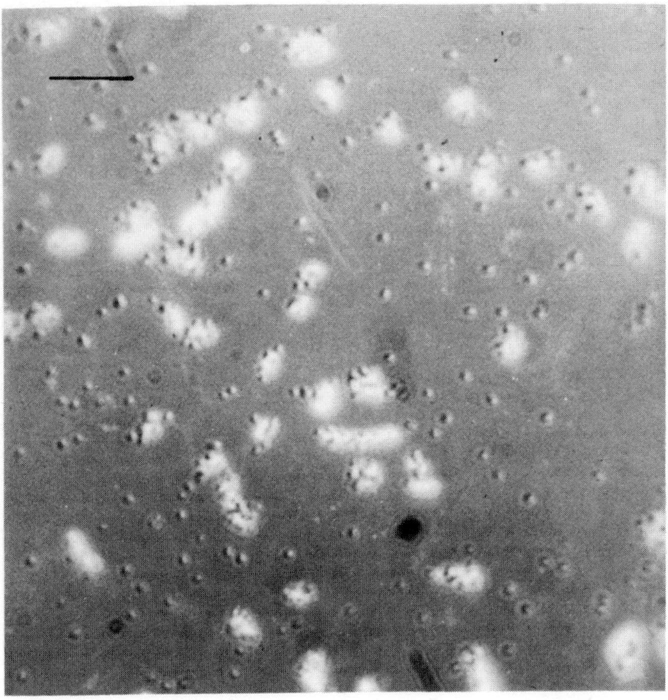

Fig. 11. Microautoradiograph of bacteria attached to glass. Although there are some randomly distributed silver grains, most are associated with the cells. Bar marker represents 5 μm.

macroorganisms. Bacteria and microalgae may be found on surfaces in layers of one to many cells thick, while the arrival of larger organisms, as for example, macroalgae and invertebrates, leads to the formation of extensive communities of attached and loosely associated organisms, often referred to as the *Aufwuchs* [Reid and Wood, 1976]. The composition of the *Aufwuchs* depends not only upon substratum characteristics but also upon the chemistry and rate of flow of the surrounding water.

I should like to gratefully thank Dr. G.I. Loeb, Naval Research Laboratories, Washington, DC, USA, for his collabor-

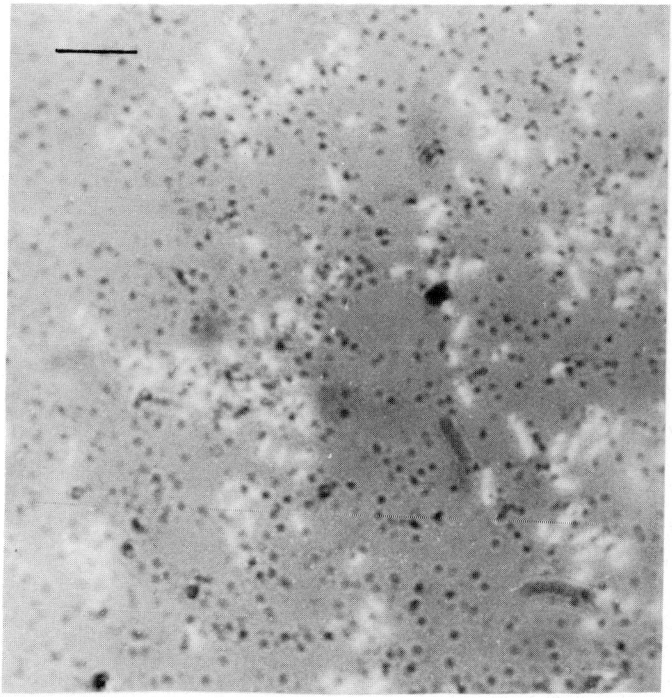

Fig. 12. Microautoradiograph of bacteria attached to polyethylene. Reduced silver grains are randomly distributed and not noticeably associated with the cells. This is apparently because the labelled material has been taken up and released by the cells by the end of the 3 h incubation period. Bar marker represents 5 μm.

ation in part of the work and for many stimulating and rewarding discussions, and Dr. L.-A. Meyer-Reil, Institut für Meereskunde, Kiel, Germany, for providing the microautoradiographic technique.

REFERENCES

Andrade, J.D. (1973). Interfacial phenomena and biomaterials. *Medical Instrumentation* 7, 110-119.

Atkinson, B. and Fowler, H.W. (1974). The significance of microbial film in fermenters. *Advances in Biochemical Engineering* 3, 221-277.

Clark, W.A. (1976). A simplified Leifson flagella stain. *Journal of Clinical Microbiology* 3, 632-634.

Corpe, W.A. (1970a). Attachment of marine bacteria to solid surfaces. In *Adhesion in Biological Systems*, pp. 73-87. Edited by R.S. Manly. New York : Academic Press.

Corpe, W.A. (1970b). An acid polysaccharide produced by a primary film-forming marine bacterium. *Developments in Industrial Microbiology* 11, 402-412.

Corpe, W.A., Matsuuchi, L. and Armbruster, B. (1976). Secretion of adhesive polymers and attachment of marine bacteria to surfaces. In *Proceedings of the Third International Biodegradation Symposium*, pp. 433-442. Edited by J.M. Sharpley and A.M. Kaplan. London : Applied Science Publishers.

Curtis, A.S.G. (1967). *The Cell Surface*. London and New York : Academic Press.

Derjaguin, B.V. and Landau, L. (1941). Theory of the stability of strongly charged lyophobic sols and of the adhesion of strongly charged particles in solutions of electrolytes. *Acta Physicochimica URSS* 14, 633-662.

Fletcher, M. (1976). The effects of proteins on bacterial attachment to polystyrene. *Journal of General Microbiology* 94, 400-404.

Fletcher, M. (1977). The effects of culture concentration and age, time and temperature on bacterial attachment to polystyrene. *Canadian Journal of Microbiology* 23, 1-6.

Fletcher, M. and Floodgate, G.D. (1973). An electron-microscopic demonstration of an acidic polysaccharide involved in the adhesion of a marine bacterium to solid surfaces. *Journal of General Microbiology* 74, 325-334.

Fletcher, M. and Floodgate, G.D. (1976). The adhesion of bacteria to solid surfaces. In *Microbial Ultrastructure: the Use of the Electron Microscope*, pp. 101-107. Edited by R. Fuller and D.W. Lovelock. London, New York and San Francisco : Academic Press.

Fletcher, M. and Loeb, G.I. (1976). The influence of substratum surface properties on the attachment of a marine bacterium. In *Colloid and Interface Science*, vol. III, pp. 459-469. Edited by M. Kerker. New York : Academic Press.

Harden, V.P. and Harris, J.O. (1953). The isoelectric point of bacterial cells. *Journal of Bacteriology* 65, 198-202.

Harwood, J.H. and Pitt, S.J. (1972). Quantitative aspects of growth of the methane oxidising bacterium *Methylococcus capsulatus* in shake flask and continuous chemostat culture. *Journal of Applied Bacteriology* 35, 597-607.

Hattori, T. and Furusaka, C. (1960). Chemical activities of *Escherichia coli* adsorbed on a resin. *Journal of Biochemistry (Japan)* 48, 831-837.

Heckels, J.E., Blackett, B., Everson, J.S. and Ward, M.F. (1976). The influence of surface charge on the attachment of *Neisseria gonorrhoeae* to human cells. *Journal of General Microbiology* 96, 359-364.

Heukelekian, H. and Heller, A. (1940). Relation between food concentration and surface for bacterial growth. *Journal of Bacteriology* 40, 547-558.

Hirsch, P. and Pankratz, St.H. (1970). Study of bacterial populations in natural environments by use of submerged electron microscope grids. *Zeitschrift fur Allgemeine Mikrobiologie* 10, 589-605.

Jannasch, H.W. (1968). Growth characteristics of heterotrophic bacteria in seawater. *Journal of Bacteriology* 95, 722-723.

Jannasch, H.W. and Pritchard, P.H. (1972). The role of inert particulate matter in the activity of aquatic microorganisms. *Memorie dell'Instituto Italiano di Idrobiologie* 29 Suppl., 289-308.

Kipling, J.J. (1965). *Adsorption from Solutions of Non-Electrolytes*. London and New York : Academic Press.

Loeb, G.I. and Neihof, R.A. (1975). Marine conditioning films. *Advances in Chemistry Series* 145, 319-335.

MacLeod, R.A. (1965). The question of the existence of specific marine bacteria. *Bacteriological Reviews* 29, 9-23.

Maroudas, N.G. (1975). Adhesion and spreading of cells on charged surfaces. *Journal of Theoretical Biology* 49, 417-424.

Marshall, K.C. (1976). *Interfaces in Microbial Ecology*. Cambridge, Mass. and London : Havard University Press.

Marshall, K.C., Stout, R. and Mitchell, R. (1971). Mechanism of the initial events in the sorption of marine bacteria to surfaces. *Journal of General Microbiology* 68, 337-348.

Mitchell, R. and Young, L. (1972). *The Role of Microorganisms in Marine Fouling*. Technical Report No. 3. US Office of Naval Research. Contract No. N00014-67-A-0298-0026 NR-306-025.

Reid, G.K. and Wood, R.O. (1976). *Ecology of Inland Waters and Estuaries*. New York : Ovan Nostrand.

Schonhorn, H. (1968). Heterogeneous nucleation of polymer melts on high-energy surfaces. II. Effect of substrate on morphology and wettability. *Macromolecules* 1, 145-151.

Shaw, D.J. (1970). *Introduction to Colloid and Surface Chemistry*. London : Butterworths.

Verwey, E.J.W. and Overbeek, J. Th.G. (1948). *Theory of the Stability of Lyophobic Colloids*. Amsterdam : Elsevier Publishing Co.

Zisman, W.A. (1964). Relation of the equilibrium contact angle to liquid and solid constituion. *Advances in Chemistry Series* 43, 1-51.

ZoBell, C.E. (1943). The effect of solid surfaces upon bacterial activity. *Journal of Bacteriology* 46, 39-59.

ZoBell, C.E. and Allen, E.C. (1935). The significance of marine bacteria in the fouling of submerged surfaces. *Journal of Bacteriology* 29, 239-251.

ZoBell, C.E. and Anderson, D.Q. (1936). Observations on the multiplication of bacteria in different volumes of stored seawater and the influence of oxygen tension and solid surfaces. *Biological Bulletin, Woods Hole* 71, 324-342.

INTERACTION OF MICROORGANISMS, THEIR SUBSTRATES AND THEIR PRODUCTS WITH SOIL SURFACES

R.G. BURNS

Biological Laboratory, University of Kent, Canterbury, Kent CT2 7NJ.

INTRODUCTION

The soil is a heterogenous milieu in which a multitude of physical, chemical and biological reactions occur. Progress in understanding these reactions has been hindered by shortcomings in methodology but, in my opinion, the major barrier has been the lack of a multi-disciplinary approach. Only when the researcher is conversant with such varied subjects as microbiology, enzymology, chemistry, hydrology, geology and colloid science can he or she hope to describe the soil environment with precision. Not surprisingly, therefore, the gaps in our knowledge remain sizeable. Fortunately new techniques are being developed and established ones critically re-viewed [Pochon, Tardieux and d'Aguilar, 1969; Roswell, 1973] and a number of articles have appeared in recent years which have assembled the large and diverse literature and seriously attempted to present a cohesive view of soil microbiology [Stotzky, 1974; Hattori and Hattori, 1976]. This article owes a considerable debt to these illuminating manuscripts as it does to the four volumes of Soil Biochemistry [McLaren and Peterson, 1967; McLaren and Skujins, 1971; Paul and McLaren 1975a, 1975b].

It is readily apparent that the features revealed by examining gram quantities of soil give, at best, an oversimplified impression of the environment in which the microbe functions. At worst, such characteristics as percent organic matter, slurry pH, water content, and oxygen level, can be totally misleading because it is the nature of colloids, molecules and ions that is the dominant influence upon microbial activity. Therefore the reactions of microorganisms at the soil colloid/soil water interface are of crucial importance in soil microbiology, for it is here that physical, chemical and biological phenomena tend to concentrate water, substrates, metabolites, enzymes and inorganic nutrients as well as the microbes themselves. As a consequence, microorganisms perform their degradative and synthetic roles predominantly at surfaces, and not whilst suspended in the soil aqueous phase.

This review describes the physico-chemical nature of soil surfaces and microorganisms as it is relevant to their interaction. In addition, some of the probable and possible consequences of this association are discussed.

PHYSICO-CHEMISTRY OF SOIL COMPONENTS

Inorganic fraction

The soil inorganic components are somewhat arbitarily divided according to size into sand, silt and clay fractions (Table 1). For reactions at surfaces, it should be readily apparent that it is the clays which exert the dominant influence. However, the coarser particles should not be dismissed completely for they will certainly contribute to such vitally important soil properties as gas and water diffusion. It is also important to realise from the outset that the clay particles do not generally exist as discrete units. Rather they have a tendency to flocculate as well as to become coated with a thin film of organic matter. The resulting combination is usually referred to as the organo-mineral complex [Greenland, 1965a, 1965b].

In addition to their high specific surface areas clays have a second property critical to their influence on biological activity; this is their electrochemical charge. The origins and natures of these two clay characteristics will be discussed only briefly here and the reader is referred to expositions by Marshall [1964], Miller, Turk and Foth [1965] and Grim [1968] for further details.

Clay minerals are comprised of two basic arrangements of molecules: silicon oxide tetrahedra and aluminium oxide or hydroxide octahedra. These primary units are assembled either in a 1:1 ratio (-Si.Al.Si.Al.Si.Al-) or a 2:1 ratio (-Si.Al.Si.Si.Al.Si-). The layers of the 1:1 silicates are tightly held together by hydrogen bonds whilst many of the 2:1 silicates are loosely associated by physical forces of the van der Waals type. Thus 2:1 clays (for example, montmorillonite) will expand upon wetting and shrink upon drying whilst 1:1 clays, such as kaolinite, are stable by comparison. Therefore expanding lattice clays have both an internal and an external surface area available for adsorption of water, nutrient ions and even some macromolecules.

Some 2:1 clays have layers which are loosely held together by magnesium ions and are only partially expanding whilst in

Table 1.
Characteristics of soil inorganic components

Soil fraction	Diameter (mm)	Number of particles (g^{-1})	Surface area ($m^2.g^{-1}$)
Fine gravel	2.00-1.00	90	0.11
Coarse sand	1.00-0.50	720	0.23
Medium sand	0.50-0.25	58×10^2	0.45
Fine sand	0.25-0.10	46×10^3	0.91
Very fine sand	0.10-0.05	72×10^4	2.27
Silt	0.05-0.002	58×10^5	4.54
Clay	0.002	90×10^9	25-1000+

others the tetrahedral sheets are prevented from separating by potassium ions.

During the formation of clay minerals some of the aluminium and silicon ions are replaced by ions of lower valancies - a process known as *isomorphous substituion.* This gives rise to an abundance of unsatisfied negative charges which are then compensated by the cations in the surrounding aqueous phase. These ions (Predominantly those of Ca, Mg, H, Al, Na, K and NH_4 in arable soils) are adsorbed strongly enough to retard their passage down the soil profile during rainfall yet are still available to microorganisms and plants. In fact the surface-located ions are more or less interchangeable with equivalent ions in solution and the quantitative measure of a clay's propensity to do this is known as its *cation exchange capacity* (CEC). Monomorillonites are particularly subject to isomorphous replacement and are consequently the most reactive clay type. A number of positive (and negative) charges can arise from broken bonds on the edges of the clay crystals resulting in a small anion exchange capacity, even though this is one or two degrees of magnitude less than the CEC.

Many soils contain a mixture of clay types although soils in warm climates tend to produce vermiculite and kaolinite whilst neutral to alkaline soils favour montmorillonite formation. Illite clays are derived from river deposits in temperate zones. Table 2 lists some of the properties of these major clay types.

Table 2.

Characteristics of soil colloids
Data from Bailey and White (1970) and Webber (1972)

Clay type	Layer-ing	Swelling	Surface area ($m^2.g^{-1}$)	CEC (meq $100g^{-1}$)	Basal spac-ing (A)
Kaolinite	1:1	non-expanding	25-50	2-10	7.2
Illite	2:1	non-expanding	75-125	15-40	10.0
Vermic-ulite	2:1	part-expanding	500-700	120-200	14.0
Montmor-illonite	2:1	expanding	700-750	80-120	17.0
Organic matter	-	expanding	500-800	200-400	-

Organic Matter

Soil organic matter can be considered as a mixture of three fractions. The first is a macroscopic segment containing plant, animal and microbial debris in the early stages of decay. The anatomy of this material is often recognisable, and in organic soils (for example muck, peat) it will represent more than 50% of the soil volume. Due to the low specific surface area of this type of organic matter it is not always considered when discussing microenvironment reactions. However, as the starting point for the colloidal humus fraction its relationship with the microflora is crucial and mechanisms by which microbes are attracted to and adhere to large insoluble substrates will be briefly considered later in this chapter.

In most mineral agricultural soils approximately 10-15% of the total organic matter is still chemically-recognisable and is derived from the breakdown of the macro-organic matter described above. This humic component (fraction two) includes carbohydrates, fats, waxes, amino acids, proteins, enzymes, pigments and antibiotics, and is generally of a transient nature - either being metabolised by the microflora or being transformed and incorporated into the third fraction (see later).

By far the most abundant humic constituents (about 80% of the total) are those complex, aromatic and recalcitrant polymers whose structure has been the subject of extensive research in recent years [Felbeck, 1971; Schnitzer and Khan, 1972;

Haider, Martin and Filip, 1975; Flaig, Beutelspacher and Rietz, 1975]. This dark coloured, acidic material is a product of the microbial and chemical synthesis of the components of fraction two described above. Whilst a good many of the monomeric components of humus have been identified (albeit following harsh chemical treatment) they do not necessarily reveal the nature of the parent polymer. The situation is further complicated by variability in the composition of humus and the ongoing processes of condensation and depolymerisation.

It is the amorphous humic fraction that has important physical and chemical properties with regard to microbial activity (Table 2). The polymers are flexible, undergo swelling and shrinking, and have significant internal surface areas. In addition, they usually carry a high proportion of negative charges arising from the pH-mediated dissociation of their constituent molecules. Therefore the isoelectric points of the humic polymers and the soil pH will determine their overall charge. The cation exchange capacity of soil organic matter is largely attributable to carboxyl groups at low pH whilst phenolic hydroxyl groups increase in importance as the pH rises. Alcoholic hydroxyl, carbonyl and methoxy moieties have also been demonstrated.

As mentioned in the previous section clay and humic colloids tend to become intimately associated to form the organo-mineral complex and it is likely that only a proportion of the silicate surface is freely available for ions and molecules. Organic nitrogenous residues may also accumulate on silt surfaces [Ladd, Parsons and Amato, 1977a, 1977b].

Microorganisms

The basic components of bacterial cell walls have been discussed by many [Glauert and Thornley, 1969; Rogers, 1970; Reavely and Burge, 1972; Meadows, 1974] and their functions speculated upon [James, 1972; Marshall, 1976]. Gram-positive cells are structurally the less complicated of the two principal types and are composed primarily of teichoic acids, proteins, polysaccharides and mucopeptides. The teichoic acids, which are polymers of glycerol or ribitol phosphate, can be replaced by teichuronic acid when phosphate is limiting [Ellwood and Tempest, 1972]. Both teichoic and teichuronic acid tend to be concentrated towards the outer surface of the cell wall and it is therefore their properties which primarily influence the way in which the cell reacts with surfaces. Gram-negative bacterial cell walls are composed of a complex multi-layered arrangement of lipo-polysaccharides, phospholipids, proteins and mucopeptides.

The charge of a cell is due to its surface ionogenic groups which undergo dissociation depending on the pH of the immediate environment. The ionisation of carboxyl and amino groups (Equations 1 and 2) is a critical reaction and indicates a net positivity in highly acid conditions and a net negativity in alkaline conditions. The dissociation constants (pK) of the wall components and their relative proportions will determine the pH at which the shift from cationic through neutral to anionic occurs. Sulphate and phosphate surface groups respond similarly to carboxyl groups.

$$\underset{\text{acid pH}}{-COOH} \longleftrightarrow \underset{\text{alkaline pH}}{-COO^- + H^+} \qquad (1)$$

$$\underset{\text{acid pH}}{-NH_3^+} \longleftrightarrow \underset{\text{alkaline pH}}{-NH_2 + H^+} \qquad (2)$$

Information about the surface ionogenic groups of bacteria can be derived from measuring the electrophoretic moblity of bacteria at various pH values (Table 3).

Table 3.
Electrophoretic mobility of microbes and ions at pH 7.0. From James (1972)

	Mobility ($m^2.s^{-1}.V^{-1} x\ 10^8$)
Klebsiella aerogenes	-1.74
Staphylococcus aureus	-1.00
Streptococcus pyogenes	-1.03
Chlorella sp.	-1.70
Hydrogen ion	+36.7
Chloride ion	-6.8
Sodium ion	+5.2

Both teichoic and teichuronic acid polymers are anionic, the charge of the former being derived from phosphate ester groupings; that of the latter from the carboxyl groups of the uronic acid moieties. The magnitude of this negativity is regulated by alanyl ester residues [Archibald, Baddiley and

Heptinstall, 1973] and, possibly, by conformational changes in the teichoic acid. The amphoteric nature of proteins allows Gram-negative bacteria to carry either a net negative or a net positive charge, arising from the ionisation of carboxyl and amino groups respectively. The component amino acids will effect the degree of response (for example, histidine has a different pK to alanine). At its isoelectric point the surface charge of a bacterium due to the proteins will be zero.

In an oft-quoted piece of work, Marshall [1967] examined the electrophoretic mobility of two species of *Rhizobium* and discovered that one (*Rhizobium lupini)* displayed a constant negative charge between pH 4.0 and 10.7 but a decrease to zero charge below pH 4.0. *Rhizobium trifolii,* in contrast, had a constant negative charge between pH 4.0 and 9.0 which then increased with increasing pH. At highly acid pH values this species became cationic.

The following factors were invoked to explain the difference between the two *Rhizobium* species. The surface carboxyl groups of *Rhizobium lupini* dissociate at pH 4.0 whilst below that (about pH 2.0) they are undissociated and as a result the cell has no surface charge. *Rhizobium trifolii* cells have a net negative charge at pH 4.0 - 9.0 due to an *excess* of acidic groups whilst at pH 2.0 some basic amino groups dissociate to give the cell a positive charge. At highly alkaline reactions none of the amino groups dissociate and thus the proportion of negative charges increases.

ASSOCIATION OF SOIL COMPONENTS WITH SURFACES

Ions

The negatively charged soil colloids attempt to establish their electro-neutrality by attracting counter-ions from the adjacent aqueous phase. As a consequence there is, according to one theory, a layer of cations (the Stern layer) firmly held to the colloid surface by electrostatic and van der Waals forces, as well as a gradient of ions responding to opposing electric attraction and thermal dispersion forces (the electrical or diffuse double layer - DDL). The Gouy-Chapman model, in contrast, does not recognise the existence of a stable layer of cations - only the DDL. If the aqueous phase associated with the colloid is substantial then the equilibrium distribution of ions within it is finite and there is an area distant from the surface which is ionically independent. In these circumstances the ionic distribution complies with the Boltzmann equation (3) where

$$C_1 \;/\; C_2 = \exp(-\Delta E/kT) \qquad (3)$$

ΔE ($= E_1 - E_2$) is the difference in potential energy of the ions comparing arbitrary positions 1 and 2 whilst kT is the (thermal) kinetic energy of the ion.

Realistically, however, the soil water level is often depleted, at which point it compresses the cloud of ions and a truncated DDL forms, a zone that tends to absorb water up to the theoretical distribution of its component ions. The thickness of the DDL, which can be calculated from the Gouy-Chapman relationships, lies in the range 2 - 40 nm and is determined by a combination of colloid type, counter-ion valence, salt concentration in the solution, and moisture level.

Anions tend to be expelled from the DDL and any adsorption that occurs is restricted to the exposed edges of the clay minerals. It is a process sensitive to pH and electrolyte level and a degree of competition is noticeable. For example, phosphate ions are adsorbed to the exclusion of sulphate ions which, in turn, are adsorbed in preference to nitrate. Fixation of phosphate can be strong enough to induce a negative charge indicating that the binding mechanism is of a chemical rather than an electrostatic nature. Iron and aluminium oxides also retain anions.

Detailed accounts of cation and anion adsorption in soil can be discovered in articles by Laudelout [1970], Mott [1970], Bolt [1976] and Bolt, Bruggenwert and Kamphorst [1976]. The value of adsorption isotherms in elucidating the nature of adsorbent-adsorbate reactions has been discussed by Giles [1970].

Substrates

Much of the research concerning the adsorption of organic substrates has involved clay surfaces [Greenland, 1970]. This is somewhat inevitable because clays, when compared to soil organic matter, have consistent and well understood chemical and physical properties.

Although many of the microbial substrates investigated are of natural origin (for example, glucose, urea, alcohols) a good deal of the emphasis has been towards pesticides. This is expedient because the availability of pesticides to plant roots, soil animals and microorganisms is of agronomic and ecological significance. As a result, an extensive volume of research, concerning the interaction of pesticides with soil colloids, has appeared in the last ten years [Guenzi, 1974]. Much of this has been related to their availability for microbial decay and the compounds involved can thus be regarded as substrates, independent of their pesticidal capacity. Of course, the size and water solubility of a pesticide will determine its mode of interaction with a soil particle, as will its ionic and/or molecular properties. Some of these rules will apply to non-pesticide substrates, ranging from the high molecular weight insoluble polymers such as cellulose, starch, and chitin to the more soluble amino acids and disaccharides.

Adsorptive mechanisms relevant to pesticides and soil have been reviewed elsewhere and the reader is referred to articles by Bailey and White [1970], Hayes [1970], Hamaker and Thompson [1972], and Weber [1972] for detailed information. Table 4 summarises the types of adsorption that can occur between pesticides and soil components and dramatically dispels the notion that sorption at colloid surfaces is restricted to those cationic moieties that can become involved in ion-exchange. The value of infrared spectroscopy and X-ray diffraction in revealing this immense range of adsorptive mechanisms has been discussed by Mortland [1970]. Only the principal modes of adsorption are outlined here.

It is important to remember, when dealing with the adsorption of substrates to surfaces, that surfaces may already be occupied by water molecules. Even air-dried soils are normally in equilibrium with a relative humidity of close to 100%.

Table 4.

Mechanisms by which microbes, enzymes, substrates, products and inorganic ions become associated with soil surfaces

SORPTION	cation exchange anion exchange protonation hemisalt formation ion-dipole interactions coordination complexes hydrogen bonding van der Waals forces π-bonding entropy effects covalent bonding.
MICROBIAL	exopolymers
STRUCTURES	flagella pili prostheca
MISCELLANEOUS	chemical incorporation into humus physical entrappment within humus hydrophobic/lipophilic reactions absorption by cells.

Ion exchange

Ion exchange involves the replacement of loosely held ions on the soil colloid surface by like-charged molecules in solution. The size of the adsorbate molecule presents problems not encountered when discussing inorganic ion adsorption, particularly with reference to the molecular weight, structure and configuration of the molecule. In practice, this type of adsorption occurs when a potential adsorbate is present as a cation (for example, the bipyridylium herbicides) or as a base capable of becoming positively charged when protonated (for example, the triazines). The sources of protons include hydrogen ions at the colloid surface, water, and any transferred from other cationic species. Susceptibility to protonation is itself related to the hydrogen ion concentration and the dissociation constants of the interactants whilst subsequent adsorption may be influenced by competition from inorganic ions.

Acidic pesticides dissociate as the hydrogen ion concentration decreases to produce anions which are likely to be repelled by the negatively-charged soil surfaces.

Hydrogen bonding

Hydrogen bonding includes a method of adsorption which involves the linking of the functional group of the organic molecule to the polarised hydration water of the exchange cation: a process known as water-bridging. Hydrogen bonds can also form between the pesticide and another adsorbed organic molecule on the colloid surface, as they can between the pesticide and the oxygen and hydroxyl groups on the clay or the carboxyl, amino, and hydroxyl groups on organic matter.

Cation-dipole interactions

This is a direct type of adsorption involving on one hand polar, non-ionic pesticides as well as some natural organic substrates [Farmer and Ahlrich, 1969] and, on the other, an exchange surface cation. In practice, ion-dipole reactions may only prove important in dehydrated microsites where competition provided by water molecules is restricted [Green, 1974].

Van der Walls forces

Van der Walls forces are short range, relatively weak physical forces often functioning in conjunction with other surface mechanisms and contributing to the total adsorption of many organic compounds. Their importance is certainly related to the size of the molecule together with its proximity to, and the number of points of contact with, the adsorbent (forces tend to be additive). They are a factor in the adsorption of uncharged non-polar substrates, such as dextran [Olness and Clapp, 1975].

Miscellaneous

Multivalent exchange cations at the clay surface can form a bridge by which electronegative atoms are attached. π-Bonding has been observed to occur between transition metals on clay surfaces and a variety of neutral molecules, such as toluene and benzene [Doner and Mortland, 1970]. Finally ligand exchange, co-ordination bonds, and entropy changes may all effect adsorption.

Microorganisms

Although microorganisms have been described as functioning at solid-liquid interfaces this should not be interpreted as meaning that they form a compact and irreversibly adsorbed layer. Marshall, Stout and Mitchell [1971] have defined two levels of cellular adsorption to solids - one reversible and the other irreversible. The former category includes those cells loosely held at or in the region of the soil surface; the latter are tightly retained stationary cells. Comparisons with the two-phase ionic associations referred to earlier, that is the Stern layer and the diffuse electrical layer, are obvious. The point is that the three-dimensional influence of soil particulates on microorganisms is both asymmetric and extensive.

The primary theoretical problem that microbes and soil particles must solve, prior to association, is their similarity of charge. No doubt the hydrophobicity of bacteria tends to propel them towards surfaces [Marshall and Cruickshank, 1973], but the anionic nature of soil colloids suggests that electrostatic forces are of minimal importance in bringing about association. Bacteria are usually positively charged below their isoelectric point (about pH 2-3) and the pH values of soil are assumed to be in excess of this. However, it is prudent to remember that microenvironment variations in hydrogen ion concentration, ion concentration and type may change the surface ionogenic properties of the reactants. Certainly experiments using ion exchange resins have shown that microorganisms can be anionic, cationic or neutral according to the hydrogen ion concentration and species [Daniels, 1972] and this knowledge has been used for resolving binary mixtures of oppositely charged cells [Daniels and Kempe, 1967].

Hattori [1973] has described some of the electrochemical processes which could be involved in binding cells to clay surfaces. These included interactions between negative bacterial surfaces and the positively charged edges of clay platelets; positively charged bacterial surfaces at acid pH values and negatively charged clay surfaces; carboxyl groups in the cell wall and the di- or polyvalent cations held at the clay face, and bacterial amino groups and the negatively charged soil surface. When an organic surface is involved a number of organic-organic electrostatic interactions are possible depending upon the nature of the humic material.

Many microorganisms produce exopolymers [Sutherland, 1972] which will anchor them firmly to solid surfaces [Harris and Mitchell, 1973] as well as compounds which may help the cell to overcome electrostatic repulsion. Substrate levels will influence the production of microbial gums [Guggenheim and Schroeder, 1967] and their formation may even be stimulated after the initial contact has been made [Fletcher and Floodgate, 1973]. Microorganisms subsequently sheared from surfaces often leave behind residual polymer material [Marshall *et al.*, 1971] - no doubt a welcome substrate for adjacent microbes [Larock and Ehrlich, 1975].

Microbes can also become attached to and embedded in the exopolysaccharides of other adsorbed microbes (important in microbe-microbe aggregation) and immersed in plant root mucigel [Greaves and Darbyshire, 1972]. Other possible functions of slime layers, especially in relation to adsorbed microbes, include imparting resistance to desiccation; causing cell orientation due to the hydrophic nature of some capsules; acting as a food source when ambient levels are low; containing exoenzymes concerned with polymer decay, and decreasing susceptibility to predators. The importance of cellular projections, such as adhesive stalks and fibrillar appendages, have been discussed by Paerl [1975].

The reverse of the phenomenon described above, the adsorption of individual clay platelets to microbial cells, has been looked at by Lehav [1962] and Marshall [1969]. They suggested that the orientation of clay particles at surfaces can be extremely variable as a result of the clay's contrasting edge and surface charges, its exchangeable ions, and the carboxyl-group-amino group ratio, influenced by the hydrogen ion concentration, at the microbial surface.

The term 'mutual adsorption' is usually applied to like-sized reactants as, for example, small clay aggregates and individual bacteria.

CONSEQUENCES OF MICROBE-SURFACE INTERACTIONS

Inorganic nutrients

A microorganism adhering to soil particulate or aggregate surfaces will be immersed in an atmosphere of ions, accumulated by the adsorbent as well as by the cell itself (Fig.1). The availability and therefore nutritional value of these ions

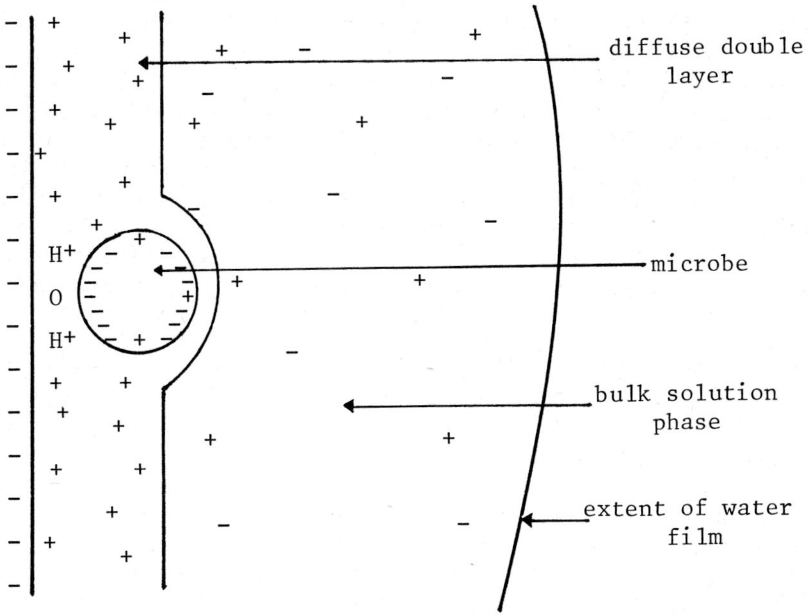

Fig. 1. The ionic environment of a soil microorganism.

is determined by their sign and valency, the extent and tenacity of their adsorption, the dimensions of the water film, together with the electrochemical properties of the microbe and its production of exchange ions.

If microorganisms can exchange ions directly with associated soil surfaces their competitive edge over non-adsorbed microbes is clear to see. If, however, desorption is a pre-requirement for ion uptake then at least a proportion of the advantage would be lost. This is because there would be a tendency for ions to rapidly equilibrate (as much as the DDL would allow) and hence be diluted in the soil water. The subsequent uptake of inorganic (and organic) nutrients may be driven by proton gradients formed by H^+ excretion, in the same way that intracellular transport is supposed to occur [Mitchell, 1967]. Humic acids may also have an important role in ion uptake analagous to that suggested for plants [Vaughan and MacDonald, 1976].

The availability of inorganic nutrients is influenced in another more direct way. This is a solubilisation process, mediated by the organic (for example, formic, citric, 2-oxoglutaric) and inorganic (for example, H_2CO_3, H_2SO_4) acids produced during microbial metabolism, and is a change in nutrient levels not merely due to corrosion but also to subtle shifts in the hydrogen ion concentration and chelation. The type and local concentration of inorganic ions (usually cations), in addition to any direct function as nutrients, may also control membrane composition and integrity, cell permeability, toxin production, cell division, spore germination, effectiveness of bacteriophage attachment and quite obviously the success or failure of lithotrophs.

Organic nutrients

The adsorption and subsequent availability of organic substrates in soil is affected by such factors as clay type, organic matter content, ionic composition and state of hydration, in addition to the physical and chemical nature of the substrate involved. The literature is full of apparently contradictory reports which chronicle the stimulation and retardation of substrate breakdown and microbial proliferation in soil. This conflict is partially resolved when one considers the two theoretical extremes: the first that the substrate is adsorbed adjacent to a degrader microbe, the second that sorption spatially separates the reactants.

Sugars are poorly adsorbed by soil clays yet Stotzky [1966], Stotzky and Rem [1967] and Novâková [1972a, 1972b] reported that non-expanding clays had little influence on the bacterial breakdown of glucose whereas lattice clays had a dramatic stimulatory effect. It was suggested that the success of microbes in glucose-amended montmorillonite was related, not to the adsorption of substrate or microbes, but to the clays ability to remove the products of metabolism and maintain a constant hydrogen ion concentration. Martin [1976] described the accelerated growth, carbon dioxide evolution and glucose consumption of actinomycetes in the presence of Ca-montmorillonite and humic acid. They also believed that the response was due to adsorption of inhibitory metabolites but not, however, to the stabilisation of the hydrogen ion concentration.

The glucose polymer, starch, is sorbed by clay minerals and its depolymerisation is accelerated by montmorillonite but not kaolinite [Filip, 1973]. Montmorillonite also stimulates the microbial decay of aldehydes [Kunc and Stotzky, 1970] yet adsorbs and inhibits dextran breakdown [Olness and Clapp, 1972, 1975] and stabilises amino acids [Sorensen, 1972, 1975]. According to Filip [1973] the fungal synthesis of humic acids is enhanced by montmorillonite.

The degradation of complex organic materials confronts the microbe with at least two major problems. The first is the location of and migration to the substrate (see Chet and Mitchell, 1976) whilst the second involves the efficient utilisation of the products of breakdown. Cellulose is a good example of a large, insoluble substrate that is physically associated with a range of polymers (uronides, pectins, lignins) and which requires a consortium of microbes and enzymes to effect its total decay [Reese, 1976]. It is not until the dimer or even the monomer stage is reached that microbes can absorb the sugar. This seems a conceptually inefficient way of getting carbon and energy and some cellulytic microorganisms (for example, *Cytophaga*) improve their chances of success by aligning their cells horizontal to the cellulose fibrils. The effect of this is to maximise cell-substrate contact, obviate the need for diffusible exoenzymes, and ensure the cell of a reasonable proportion of the cellobiose and glucose produced. It is also possible that microbes, when forming enzyme-substrate complexes, are using the enzymes on the cell wall for attachment as well as degradation.

It would seem that selection pressures in soil are so extreme that microorganisms which have no efficient way of utilising the products of their enzymic efforts would disappear rapidly. Possible adaptations to the problem of large extracellular substrates, which require enzymic decay prior to uptake into the cytoplasm, have been speculated upon elsewhere [Burns, 1978]. The value of colloid-bound enzymes as substitutes for continuous microbial enzyme production is outlined below.

Pesticide breakdown

The relationship of a microorganism to a pesticide is primarily one in which the agrochemical is used for energy and cell synthesis, although co-metabolism is a second degradative possibility. Non-biological mechanisms of pesticide attenuation, many of which are stimulated by adsorption to clay and organic surfaces, are plentiful. Photo- and chemical decomposition of one large group, herbicides, has recently been reviewed by Crosby [1976].

The adsorptive behaviour of a pesticide is a prime factor in determining its availability to plants, insects and, in the present context, microorganisms. Cationic pesticides, when irreversibly adsorbed within the interlammelar speces of montmorillonite, are quite obviously protected from a direct attack by microbes (d (001) spacing = 1 - 2 nm) [Weber, 1972]. Exoenzymic breakdown also appears unlikely. Paraquat and diquat adsorbed on the external surface of clay [Weber and Coble, 1968] or organic matter [Burns and Audus, 1970] are somewhat more mobile and even subject to microbial decay.

The availability of basic pesticides (for example, *sym*-triazines) to degradative microbes is related to the nature of the adsorbant, the level of hydration and the hydrogen ion concentration. A decline in the latter generally increases adsorption (depending on the pK_a of the pesticide) although molecular structures can also be adsorbed. Interlammelar adsorption of triazine herbicides occurs but is not irreversible. The consequences of intercalation for triazine persistence is not at all clear from the literature (see review by Weber, 1970). Generally breakdown is retarded but there are many instances of stimulation. In reports of the latter event it is sometimes difficult to distinguish between microbial and chemical decay. Yet it is probable that both mechanisms could work in tandem with a surface catalysed chemical decay converting a microbially-recalcitrant molecule to a more readily available substrate.

Acidic compounds (such as the phenoxyacetic acids) are adsorbed to some extent by organic matter but hardly at all by clays and are thus fairly mobile in soil. Much of the rapid microbial decay of this group may be occurring in the vicinity of organic surfaces where there is a high indigenous microflora [Burns and Gibson, 1977].

Many pesticides do not ionise significantly in aqueous systems. They are an extremely diverse group with a range of solubility, volatility and molecular size. Those non-ionics with a low solubility tend to be rejected from the aqueous phase and to adhere to lipophilic organic surfaces through dipole-dipole interactions. Others, which are capable of forming weak acids (phenylureas) or weak bases may utilise a wider variety of adsorptive mechanisms. More specifically the chlorinated hydrocarbons are extremely persistent in soil (half-lives measured in years) because of their structural unsuitability as a sole carbon source, their low solubility, and their propensity to complex with or even become an integral component of, humic matter. With regard to complex formation lipid-pesticide bonding is important [Pierce, Olney and Felbeck 1971]; the lipid component of humus is between one and five percent. Some chemical decay of organochlorines can occur at clay surfaces and their microbial availability is somewhat enhanced in anaerobic soils [Evans, 1977].

Organophosphorus insecticides have higher water solubilities than the chlorinated hydrocarbons and are adsorbed relatively strongly by soil particulates. Most are readily degraded in soil by a combination of chemical and biological processes (half-lives measured in days and weeks), although interlammelar adsorption by montmorillonite clays, through hydrogen bonding, may give some stability [Bowman, Adams and Fenton, 1970; Saltzman and Yariv, 1976]. It has been suggested [Gibson and Burns, 1977] that the primary breakdown of malathion is mediated by accumulated exoenzymes, the secondary mechanism is the function of microbial proliferation, and the tertiary route is caused by clay - surface hydrolysis [Saltzman, Yaron and Minglegrin, 1974]. All three of these events are associated with the organo-mineral complex.

The persistence of a pesticide and its breakdown products can also be modified following their incorporation into humic matter during its synthesis [Hsu and Bartha, 1973; Mather and Morley, 1975; Wolf and Martin, 1976]. In this way the life of a pesticide may be prolonged because its physical and chemical availability as a substrate is reduced. Some aromatic molecules (benzene, and possibly some pesticides) can be polymerised abiologically when adsorbed to clays [Mortland and Halloran, 1976].

Pesticides may also be absorbed by microbes, thereafter remaining unchanged in their intracellular location; a phenomenon with profound implications for biomagnification of toxic residues.

Accumulated enzymes

There is little doubt that enzymes, separated spatially and temporally from their cellular origins, accumulate in soil [Burns, 1977]. These enzymes become bound, possibly by both chemical and physical mechanisms, to the soil colloids. Leakage and replenishment of this catalytic component of soil establishes a steady-state level of activity which may contribute significantly to the breakdown of organic matter. For example, it has been estimated that between 40 and 90% of urea hydrolysis in soil is the function of an accumulated colloid-bound urease and independent of immediate microbial metabolism [McGarity and Myers, 1967; Paulson and Kurtz, 1969; Nannipieri *et al.*, 1974; Pettit *et al.*, 1976].

Currently the concept of enzymes being trapped in organic colloids is in favour and gains credence from the properties of the artificial enzyme-humic fractions produced by Ladd and his colleagues [Ladd and Butler, 1975]. In addition, the related observations of the group led by Martin at the University of California, Riverside [Verma, Martin and Haider, 1975; Verma and Martin, 1976] have underlined the ability of humus to form stable complexes with a wide variety of naturally-occurring organic molecules (including proteins). No doubt clays aid the stabilisation process by immobilising the organic colloids but they may have little direct effect on the enzyme.

Mere adsorption of enzymes on and within soil components does not offer them long-term protection from denaturation and degradation nor does it allow them to maintain activity. Therefore it seems probable that enzymes become incorporated into organic matter during humification and furthermore that the cytoplasmic and periplasmic enzymes originating from dead cells may even remain associated with cellular debris and are therefore never fully exposed to the hostile soil environment. Truly extracellular enzymes must be remarkably inefficient, surviving only for a short time on soil surfaces or needing to rapidly locate a suitable substrate. The enzyme degrading capacity of soil is enormous and most lysing microbes release proteases.

A model has been suggested to account for the persistence of at least some of the enzymes accumulated in soil [Burns, Pukite and McLaren, 1972; Pettit *et al.*, 1976]. According to this scheme enzymes become trapped inside the porous humic polymers, are protected from proteolysis and yet are accessible to substrates (Fig.2). It is possible to propose a number of advantages for a microorganism arising from its association with an enzyme-organic matter surface. The microbes are not only 'on the spot' when a suitable low molecular weight product appears but there is no need for them to expend energy synthesising the relevant exo-enzyme(s).

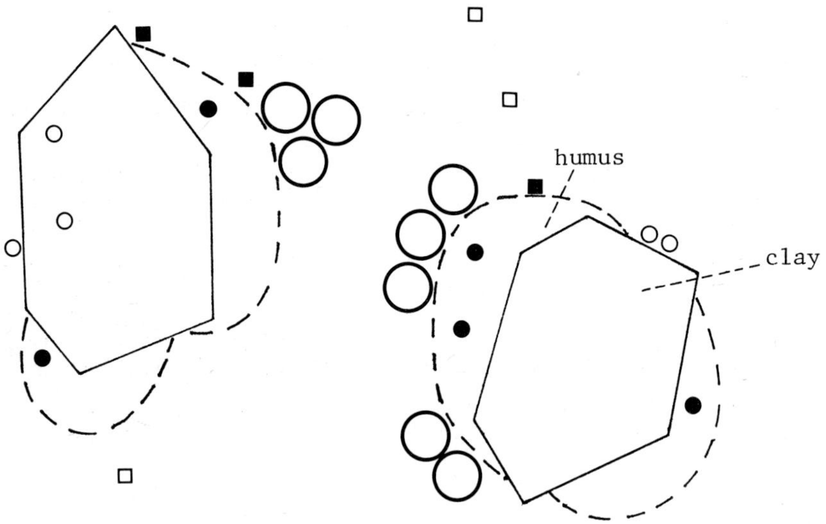

Fig. 2. Theoretical distribution of exoenzymes in the soil microenvironment. ● = *enzyme trapped within or complexed with humus;* ○ = *enzyme adsorbed to clay - either on the surface or between the lattices;* ■ = *enzyme attached to surface of organic film;* □ = *ephemeral enzyme activity in the soil aqueous phase; the larger open circles represent microorganisms.*

The accumulated enzyme fraction may also be important in the initial response to a substrate in advance of microbial enrichment and/or induction. Indeed the levels of product released by accumulated enzymes may be instrumental in initiating enzyme induction in adjacent microbes, because polymers incapable of entering the cell cannot possibly bring about enzyme synthesis. The only other way a microorganism could be sure of responding to such a substrate would be to continuously produce all sorts of exoenzymes - a much less efficient way of obtaining carbon and energy. In a broader context exo-enzymes can be seen as just one factor controlling the gradual release of nutrients from soil humus; nutrients which are capable of sustaining a viable associated microflora which is then, in its turn, competent to respond to any sudden influx of organic matter.

The problems of measuring and interpreting enzyme kinetics in heterogenous soil environments are considerable [McLaren, 1960; McLaren and Packer, 1970; Cervelli *et al.*, 1973; Pettit *et al.*, 1977]. One important characteristic that does emerge from enzyme studies is that the hydrogen ion concentration at the soil surface is somewhat different from that routinely measured in slurries. The accumulation of H^+ at surfaces may account for the difference in pH (ΔpH) between the soil solution and the solid/liquid interface. Comparing the pH optimum of an enzyme in solution with its apparent (that is slurry) pH optimum in soil (see, for example, Skujins, Pukite and McLaren, 1974) reveals that the ΔpH is in the region of 1 - 2 units [McLaren and Skujins, 1968]. The use of Hammett indicators should confirm that at optimum activity the hydrogen ion concentration at the surface (where the enzyme - substrate interaction is occurring) is the same as that recorded in a soil-free solution. Disturbingly, perhaps, the reverse pH shift has been recorded and some microbial activities are carried out in soils at a lower pH than in liquid media [Weber and Gainey, 1962]. However, equally probable is that surface-adsorbed NH_4^+ could give rise to alkaline microenvironments [Williams and Mayfield, 1971] but it is clear that this aspect of microbial ecology demands further research.

Miscellaneous

It has long been known that stable soil aggregates encourage the diffusion of gases, solvents and solutes, aid root development, stimulate the production and retention of humus, and support active microbial populations. Not surprisingly, therefore, the factors which influence soil aggregate formation have attracted the attention of many workers [Baver, 1968; Aspiras *et al.*, 1971; Cheshire, 1977]. The initial stages of crumb formation may involve the flocculation of clay particles due to electrostatic and other forces [Santoro and Stotzky, 1967], and the accumulation of clay at microbial surfaces. These microaggregates may then become associated by the activities of fungal and actinomycete mycelia and possibly microbial gums [Hepper, 1975]. Subsequent stabilisation of aggregates may be brought about by aromatic compounds [Griffiths and Burns, 1972] synthesised by the associated microflora. Giovannini and Sequi [1976] envisaged the microaggregates as being held together by an organic-polyvalent ion meshwork whilst Hamblin and Greenland [1977] underlined the importance of Fe- and Al-organic matter in stabilisation.

The range of microsites within a soil aggregate, where there is a marked discontinuity of oxygen, water and nutrients, is immense [Tyagny-Ryadno, 1968; Nishio, 1970]. Therefore microorganisms located at the centre of aggregates will be living in a very different environment to those cells on the periphery. No doubt those microorganisms (or their progeny) involved in building aggregates become trapped within them and are physically restricted in a zone which, although low in oxygen, protects them from predators and the immediate effects of dehydration. The availability of inorganic and organic nutrients is probably altered when particles are aggregated and data relevant to discrete colloid environments may need some re-interpretation.

A series of publications by Stotzky and his colleagues [Stotzky *et al.*, 1961; Stotzky and Martin, 1963] describe how the presence of montmorillonite clays may lower the incidence of banana wilt *(Fusarium oxysporum f.cubense)* - a result not induced by other soil components. A similar textural relationship was suggested to account for the discontinous geographical distribution of *Histoplasma capsulatum* [Stotzky and Post, 1967]. That these effects are a direct result of the sorption of fungal cells and thus the restricted dispersion of propagules is unlikely. What is more probable is that the expanding lattice clay supports a type of biochemical and microbiological environment which is unsuitable for the establishment and success of certain invading fungal species.

The adsorption of clay platelets onto a microorganism's surface may protect the enveloped cell from predators and parasites [Roper and Marshall, 1974]. Similarly montmorillonite will increase the survival rate of fast-growing rhizobia under dehydrating conditions by retaining water in their lattices [Bushby and Marshall, 1977]. Enhancement of resistance to high temperatures and X-rays has been reported for microbe-montmorillonite mixtures [Marshall, 1964; Muller and Schmidt, 1966].

Table 5.

Some properties of soil colloids which influence microbial activity

CLAY	ORGANIC
large unit surface area	large unit surface area
concentrates and exchanges inorganic nutrients	concentrates and exchanges inorganic nutrients
concentrates and exchanges organic nutrients	concentrates and exchanges organic nutrients
retains water	retains water
catalyses non-biological hydrolysis	catalyses non-biological hydrolysis
involved in formation of aggregate environment	involved in formation of aggregate environment
acts as buffer (H^+ adsorption)	source of organic nutrient *per se*
adsorbs toxic metabolites	stabilises exoenzymes
adsorbs antibiotics	adsorbs lipophilic substrates
immobilises organic cations	aids ion absorption
catalyses non-biological synthesis	has bacteriostatic properties
physically protects microbes	incorporates substrates
immobilises phage particles	stimulates chemotaxis

CONCLUSION

It is hard to imagine a more complex environment for microbial activity than soil, in which the interplay of chemical, physical and biological forces is apparently infinite. It should be obvious from this exposition that soil microbes do not lead independent lives, instead they interact at the molecular, colloidal and aggregate level. These interactions are reciprocal and it is axiomatic that the environment selects the microbe and the microbe creates the environment. This suggests, in ecological terminology, a sequence of colonisation and succession in microbial communities, followed by a climax state held in dynamic equilibrium by a myriad of self-regulatory mechanisms. Thus when studying the microbial community in soil, there are considerations of both space and time in addition to the more obvious biological factors.

Soil surfaces play a vital and effective role in maintaining environmental equilibria even during quite dramatic perturbations such as drought, nutrient deplention, fertiliser additions and pesticide applications. This they do by buffering pH shifts, concentrating and immobilizing nutrient ions, serving as a source of slow-release organic substrates (humus), trapping water and toxic metabolites, accumulating and protecting exoenzymes, as well as masking microorganisms from the effects of dehydration and predation (Table 5 page 129).

It is unlikely that all, or even many, of these interwoven relationships will ever be untangled and fully understood. Nevertheless, the study of microbial reactions at surfaces perhaps provides the most promising avenue for future research into the fundamental problems of soil microbiology and biochemistry.

REFERENCES

Archibald, A.R., Baddiley, J. and Heptinstall, S. (1973). The alanine ester content and magnesium binding capacity of walls of *Streptococcus aureus* H grown at different pH values. *Biochimica et Biophysica Acta* 291, 629-634.

Aspiras, R.B., Allen, O.N., Harris, R.F. and Chesters, G. (1971). The role of microorganisms in the stabilisation of soil aggregates. *Soil Biology and Biochemistry* 3, 347-353.

Bailey, G.W., and White, J.L. (1970). Factors influencing the adsorption, desorption and movement of pesticides in soil. *Residue Reviews* 32, 29-92.

Baver, L.D. (1968). The effect of organic matter on soil structure. *Pontificia Academia Scientarium Scripta Varia* 32, 1-31.

Bolt, G.H. (1976). Adsorption of anions by soil. In *Soil Chemistry A. Basic Elements*, pp. 91-95. Edited by G.H. Bolt, and M.G.M. Bruggenwert. Amsterdam, Oxford and New York : Elsevier.

Bolt, G.H., Bruggenwert, M.G.M. and Kamphorst, A. (1976). Adsorption of cations by soil. In *Soil Chemistry A. Basic*

Elements, pp. 54-90. Edited by G.H. Bolt and M.G.M. Bruggenwert. Amsterdam, Oxford and New York : Elsevier.

Bowman, B.T., Adams, R.S. and Fenton, S.W. (1970). Effect of water upon malathion adsorption onto five montmorillonite systems. *Journal of Agriculture and Food Chemistry* 18, 723-727.

Burns, R.G. (1977). Soil enzymology. *Science Progress (Oxford)* 64, 281-291.

Burns, R.G. (1978). Enzyme activity in soil: some theoretical and practical considerations. In *Soil Enzymes*. Edited by R.G. Burns. London and New York : Academic Press.

Burns, R.G. and Audus, L.J. (1970). Distribution and breakdown of paraquat in soil. *Weed Research* 10, 49-58.

Burns, R.G. and Gibson, W.P. (1979). The disappearance of 2, 4-D, diallate and malathion from soil and soil components. In *Agrochemicals in Soils*. Edited by A. Banin. New York and Berlin : Springer-Verlag (in the press).

Burns, R.G., Pukite, A.H. and McLaren, A.D. (1972). Concerning the location and persistence of soil urease. *Soil Science Society of America Proceedings* 36, 308-311.

Bushby, H.V.A. and Marshall, K.C. (1977). Some factors affecting the survival of root-nodule bacteira on dessication. *Soil Biology and Biochemistry* 9, 143-147.

Cervelli, S., Nannipieri, P., Ceccanti, B. and Sequi, P. (1973). Michaelis constant of soil acid phosphatase. *Soil Biology and Biochemistry* 5, 841-845.

Cheshire, M.V. (1977). Origins and stability of soil polysaccharides. *Journal of Soil Science* 28, 1-10.

Chet, I. and Mitchell, R. (1976). Ecological aspects of microbial chemotactic behaviour. *Annual Review of Microbiology* 30, 221-239.

Crosby, D.G. (1976). Nonbiological degradation of herbicides in soil. In *Herbicides: Physiology, Biochemistry, Ecology* vol. 2, pp. 65-97. Edited by L.J. Audus. London and New York : Academic Press.

Daniels, S.L. (1972). The adsorption of microorganisms onto solid surfaces: a review. *Developments in Industrial Microbiology* 13, 211-253.

Daniels, S.L. and Kempe, L.L. (1967). The separation of bacteria by adsorption onto ion exchange resins. II. Resolution of binary mixtures. In *Chemical Engineering in Biology and Medicine*, pp. 391-415. Edited by D. Hershey. New York : Plenum.

Doner, H.E. and Mortland, M.M. (1970). Benzene complexes with copper (II) montmorillonite. *Science* 166, 1406-1407.

Ellwood, D.C. and Tempest, D.W. (1972). Effects of environment on bacterial wall content and composition. *Advances in Microbial Physiology* 7, 83-117.

Evans, W.C. (1977). Biochemistry of the bacterial catabolism of aromatic compounds in anaerobic environments. *Nature, London* 270, 17-22.

Farmer, W.J. and Ahlrich, J.L. (1969). Infrared studies of the mechanism of adsorption of urea-d_4, methylurea-d_3 and 1,1-dimethylurea-d_2 by montmorillonite. *Soil Science Society of America Proceedings* 33, 254-258.

Felbeck, G.T. (1971). Chemical and biological characterisation of humic matter. In *Soil Biochemistry* vol. 2, pp. 36-59. Edited by A.D. McLaren, and J. Skujins. New York : Marcel Dekker.

Filip, Z. (1973). Clay minerals as a factor influencing the biochemical activity of soil microorganisms. *Folia Microbiologica* 18, 56-74.

Flaig, W., Beutelsphacher, H. and Rietz, E. (1975). Chemical composition and physical properties of humic substances. In *Soil Components* vol. 1, pp. 1-211. Edited by J.E. Gieseking. Berlin : Springer-Verlag.

Fletcher, M. and Floodgate, G.D. (1973). An electron-microscopic demonstration of an acidic polysaccharide involved in the adhesion of a marine bacterium to solid surfaces. *Journal of General Microbiology* 74, 325-334.

Gibson, W.P. and Burns, R.G. (1977). Breakdown of malathion in soil and soil components. *Microbial Ecology* 3, 219-230.

Giles, C.H. (1970). Interpretation and use of sorption isotherms. In *Sorption and Transport Processes in Soil.* SCI Monograph 37, pp. 14-32.

Giovannini, G. and Sequi, P. (1976). Iron and aluminium as cementing substances of soil aggregates. I and II. *Journal of Soil Science* 27, 140-153.

Glauert, A.M. and Thornley, M.J. (1969). The topography of the bacterial cell wall. *Annual Review of Microbiology* 23, 159-198.

Greaves, M.P. and Darbyshire, J.F. (1972). The ultrastructure of the mucilaginous layer on plant roots. *Soil Biology and Biochemistry* 4, 443-449.

Green, R.E. (1974). Pesticide-clay-water interactions. In *Pesticides in Soil and Water*, pp. 3-37. Edited by W.D. Guenzi. Madison, Wisconsin : Soil Science Society of America Inc.

Greenland, D.J. (1965a). Interactions between clays and organic compounds in soils. Part II. Adsorption of soil organic compounds and its effect on soil properties. *Soils and Fertilizers* 28, 521-532.

Greenland, D.J. (1965b). Interactions between clays and organic compounds in soils. Part I. Mechanism of interaction between clays and defined compounds. *Soils and Fertilizers* 28, 415-425.

Greenland, D.J. (1970). Sorption of organic compounds by clays and soils. In *Sorption and Transport Processes in Soil.* SCI Monograph No. 37, pp. 79-91.

Griffiths, E. and Burns, R.G. (1972). Interaction between phenolic substances and microbial polysaccharides in soil aggregation. *Plant and Soil* 36, 599-612.

Grim, R.E. (1968). *Clay Mineralogy*, 2nd edition, p. 596. New York : McGraw-Hill.

Guenzi, W.D. (1974). *Pesticides in Soil and Water.* Madison, Wisconsin : Soil Science Society of America Inc.

Guggenheim, B. and Schroeder, H.E. (1967). Biochemical and morphological aspects of extracellular polysaccharides produced by Cariogenic streptococci (rat). *Helvetica Odontologica Acta* 11, 131-152.

Haider, K., Martin, J.P. and Filip, Z. (1975). Humus biochemistry. In *Soil Biochemistry* vol. 4, pp. 195-244. Edited by E.A. Paul and A.D. McLaren. New York : Marcel Dekker.

Hamaker, J.W. and Thompson, J.M. (1972). Adsorption. In *Organic Chemicals in the Soil Environment*, vol. 1. pp. 49-143. Edited by C.A.I. Goring and J.W. Hamaker. New York : Marcel Dekker.

Hamblin, A.P. and Greenland, D.J. (1977). Effect of organic constituents and complexed metal ions on aggregate stability of some East Anglian soils. *Journal of Soil Science* 28, 410-416.

Harris, R.H. and Mitchell, R. (1973). The role of polymers in microbial aggregation. *Annual Review of Microbiology* 27, 27-50.

Harter, R.D. and Stotzky, G. (1971). Formation of clay-protein complexes. *Soil Science Society of America Proceedings* 35, 383-389.

Hattori, T. (1973). *Microbial Life in the Soil*, p. 427. New York : Marcel Dekker.

Hattori, T. and Hattori, R. (1976). The physical environment in soil microbiology: an attempt to extend principles of microbiology to soil microorganisms. *CRC Critical Reviews of Microbiology* 4, 423-461.

Hayes, M.H.B. (1970). Adsorption of triazine herbicides on soil organic matter. *Residue Reviews* 32, 131-174.

Hepper, C.M. (1975). Extracellular polysaccharides of soil bacteria. In *Soil Microbiology*, pp. 93-110. Edited by N. Walker. London : Butterworths.

Hsu, T.S and Bartha, R. (1973). Interaction of pesticide derived chloroaniline residues with soil organic matter. *Soil Science* 116, 444-452.

James, A.M. (1972). The electrochemistry of bacterial surfaces. *Inaugural Lecture, Bedford College (Univ. London)*, October 1972. pp. 3-28.

Kunc, F. and Stotzky, G. (1970). Breakdown of some aldehydes in soils with different amounts of montmorillonite and kaolinite. *Folia Microbiologica* 15, 216.

Ladd, J.N. and Butler, J.H.S. (1975). Humus-enzyme systems and synthetic, organic polymer analogs. In *Soil Biochemistry,* vol. 4, pp. 143-194. Edited by E.A. Paul and A.D. McLaren. New York : Marcel Dekker.

Ladd, J.N., Parsons, J.W. and Amato, M. (1977a). Studies of nitrogen immobilization and mineralization in calcareous soils - II. Mineralization of immobilized nitrogen from soil fractions of different particle size and density. *Soil Biology and Biochemistry* 9, 319-325.

Ladd, J.N., Parsons, J.W. and Amato, M. (1977b). Studies of nitrogen immobilization and mineralization in calcareous soils - I. Distribution of immobilized nitrogen amongst soil fractions of different particle size and density. *Soil Biology and Biochemistry* 9, 309-318.

Larock, P.A. and Ehrlich, H.L. (1975). Observations of bacterial microcolonies on the surface of ferromanganese nodules from Blake plateau by scanning electron microscopy. *Microbial Ecology* 2, 84-96.

Laudelout, H. (1970). Cation exchange in soils. In *Sorption and Transport Processes in Soil.* SCI Monograph 37, pp. 33-39.

Lehav, N. (1962). Adsorption of sodium bentonite particles on *Bacillus subtilis. Plant and Soil* 17, 191-208.

McGarity, J.W. and Myers, M.G. (1967). A survey of urease activity in soils of northern New South Wales. *Plant and Soil* 27, 217-238.

McLaren, A.D. (1960). Enzyme action in structurally restricted systems. *Enzymologia* 21, 356-364.

McLaren, A.D. and Packer, L. (1970). Some aspects of enzyme reactions in heterogenous systems. *Advances in Enzymology* 33, 245-308.

McLaren, A.D. and Peterson, G.H. (1967). *Soil Biochemistry,* vol. 1. New York : Marcel Dekker.

McLaren, A.D. and Skujins, J. (1968). The physical environment of microorganisms in soil. In *The Ecology of Soil Bacteria.* Edited by T.R.G. Gray and D. Parkinson. Liverpool : Liverpool University Press.

McLaren, A.D. and Skujins, J. (1971). *Soil Biochemistry,* vol. 2, New York : Marcel Dekker.

Marshall, C.E. (1964). *The Physical Chemistry and Mineralogy of Soils* vol. 1. New York and London : John Wiley and Sons, Inc.

Marshall, K.C. (1964). Survival of root-nodule bacteria in dry soils exposed to higher temperatures. *Australian Journal of Biological Sciences* 20, 429-438.

Marshall, K.C. (1967). Electrophoretic properties of fast- and slow- growing species of *Rhizobium*. *Australian Journal of Biological Science* 20, 429-438.

Marshall, K.C. (1969). Orientation of clay particles sorbed on bacteria possessing different ionogenic surfaces. *Biochimica et Biophysica Acta* 193, 472-474.

Marshall, K.C. (1976). *Interfaces in Microbial Ecology*. Cambridge, Massachusetts and London, England : Harvard University Press.

Marshall, K.C. and Cruickshank, R.H. (1973). Cell surface hydrophobicity and the orientation of certain bacteria at interfaces. *Archives for Microbiology* 91, 29-40.

Marshall, K.C., Stout, R. and Mitchell, R. (1971). Mechanism of the initial events in the sorption of marine bacteria to surfaces. *Journal of General Microbiology* 68, 337-348.

Martin, J.P., Filip, Z. and Haider, K. (1976). Effect of montmorillonite and humate on growth and metabolic activity of some actinomycetes. *Soil Biology and Biochemistry* 8, 409-413.

Mather, S.P. and Morley, H.V. (1975). A biodegradation approach for investigating pesticide incorporation into soil humus. *Soil Science* 120, 238-240.

Meadows, P.W. (1974). Structure and synthesis of bacterial walls. In *Companion to Biochemistry*, pp. 343-365. Edited by A.T. Bull, J.R. Lagnado, J.O. Thomas and K.F. Tipton London : Longmans.

Miller, C.E., Turk, L.M. and Foth, H.D. (1965). *Fundamentals of Soil Science*, 4th edition. New York and London : John Wiley and Sons Inc.

Mitchell, P. (1967). Translocations through natural membranes. *Advances in Enzymology* 29, 33-87.

Mortland, M.M. (1970). Clay-organic complexes and interactions. *Advances in Agronomy* 22, 75-117.

Mortland, M.M. and Halloran, L.J. (1976). Polymerization of aromatic molecules on smectite. *Soil Science Society of America Journal* 40, 367-370.

Mott, C.J.B. (1970). Sorption of anions by soils. In *Sorption and Transport Processes in Soil* SCI Monograph 37, pp. 40-53.

Müller, H.P. and Schmidt, L. (1966). Kontinuierliche Atmungsnessungen an *Azotobacter chroococcum* Beij in Montmorillonitunter chronischer Rüntgenbestrahlung. *Archives of Microbiology* 54, 70-79.

Nannipieri, P., Ceccanti, B., Cervelli, S. and Sequi, P. (1974). Use of 0.1M pyrophosphate to extract urease from a podzol. *Soil Biology and Biochemistry* 6, 359-362.

Nishio, M. (1970). The distribution of nitrifying bacteria in soil aggregates. *Soil Science and Plant Nutrition* 16, 24-

27.
Novakova, J. (1972a). Effect of increasing concentrations of clay on the decomposition of glucose. I. Effect of bentonite. *Zentralblatt für Bakteriologie Parasitenkunde Infektionskrankheiten und Hygiene II.* Abt II. 127, 359-366.
Novakova, J. (1972b). Effect of increasing concentrations of clay on the decompsotion of glucose II. Effect of kaolinite. *Zentralblatt für Bakteriologie Parasitenkunde Infektionskrankheiten und Hygiene II.* Abt II. 127, 367-372.
Olness, A. and Clapp, C.E. (1972). Microbial degradation of a montmorillonite-dextran complex. *Soil Science Society of America Proceedings* 36, 179-181.
Olness, A. and Clapp, C.E. (1975). Influence of polysaccharide structure on dextran adsorption by montmorillonite. *Soil Biology and Biochemistry* 7, 113-118.
Paerl, H.W. (1975). Microbial attachment to particles in marine and freshwater ecosystems. *Microbial Ecology* 2, 73-83.
Paul, E.A. and McLaren, A.D. (1975a). *Soil Biochemistry* vol. 3. New York : Marcel Dekker.
Paul, E.A. and McLaren, A.D. (1975b). *Soil Biochemistry* vol. 4. New York : Marcel Dekker.
Paulson, K.N. and Kurtz, L.T. (1969). Locus of urease activity in soil. *Soil Science Society of America Proceedings* 33, 897-901.
Pettit, N.M., Gregory, L.J., Freedman, R.B. and Burns, R.G. (1977). Differential stabilities of soil enzymes : assay and properties of phosphatase and arylsulphatase. *Biochimica et Biophysica Acta* 485, 357-366.
Pettit, N.M., Smith, A.R.J., Freedman, R.B. and Burns, R.G. (1976). Soil urease : activity, stability and kinetic properties. *Soil Biology and Biochemistry* 8, 479-484.
Pierce, R.H., Olney, C.E. and Felbeck, G.T. (1971). Pesticide adsorption in soils and sediments. *Environmental Letters* 1, 157-172.
Pochon, J., Tardieux, P. and D'aguilar, J. (1969). Methodological problems in soil biology. In *Soil Biology - Reviews of Research*, pp. 13-63. Paris : UNESCO.
Reaveley, D.A. and Burge, R.E. (1972). Walls and membranes in bacteria. *Advances in Microbial Physiology* 7, 1-81.
Reese, E.T. (1976). History of the cellulase program at the U.S. Army Natrick Development Center. In *Enzymatic Conversion of Cellulosic Materials*. Edited by E. Gaden, M.M. Mandels, E.T. Reese and L.A. Spano. New York : Wiley Interscience.
Rogers, H.J. (1970). Bacterial growth and the cell envelope. *Bacteriological Reviews* 34, 194-214.
Roper, M.M. and Marshall, K.C. (1974). Modification of the interaction between *Escherichia coli* and bacteriophage in

saline sediment. *Microbial Ecology* 1, 1-13.

Roswell, T. (1973). *Modern Methods in the Study of Microbial Ecology*. Bulletin 17. Ecological Resarch Committee. Stockholm : Swedish Natural Science Research Council.

Saltzman, S. and Yariv, S. (1976). Infrared and X-ray study of parathion-montmorillonite sorption complexes. *Soil Science Society of America Journal* 40, 34-38.

Saltzman, S., Yaron, B. and Mingelgrin, U. (1974). The surface catalysed hydrolysis of parathion on kaolinite. *Soil Science Society of America Proceedings* 38, 231-234.

Santoro, T. and Stotzky, G. (1967). Influence of cations on flocculation of clay minerals as determined by the electrical sensizing zone particle analyzer. *Soil Science Society of America Proceedings* 31, 761-765.

Schnitzer, M. and Khan, S.U. (1972). *Humic Substances in the Environment*. New York : Marcel Dekker.

Skujins, J., Pukite, A. and McLaren, A.D. (1974). Adsorption and activity of clutinase on kaolinite. *Soil Biology and Biochemistry* 6, 179-182.

Sorensen, O.H. (1972). Stabilization of newly-formed amino acid metabolites in soil by clay minerals. *Soil Science* 114, 5-11.

Sorensen, L.H. (1975). The influence of clay on the rate of decay of amino acid metabolites synthesised in soils during decomposition of cellulose. *Soil Biology and Biochemistry* 7, 171-177.

Stotzky, G. (1966). Influence of clay minerals on microorganisms: III. Effect of particle size, cation exchange capacity and surface area on bacteria. *Canadian Journal of Microbiology* 12, 1235-1245.

Stotzky, G. (1974). Activity, ecology, and population dynamics of microorganisms in soil. In *Microbial Ecology*, pp. 57-135. Edited by A. Laskin, and H. Lechevalier. Cleveland, Ohio : CRC Press.

Stotzky, G., Dawson, J.E., Martin, R.T. and Ter Kuile, G.H.H. (1961). Soil mineralogy as a factor in the spread of *Fusarium* wilt of banana. *Science* 133, 1483-1484.

Stotzky, G. and Martin, R.T. (1963). Soil mineralogy in relation to the spread of *Fusarium* wilt of banana in Central America. *Plant and Soil* 18, 317-321.

Stotzky, G. and Post, A.H. (1967). Soil mineralogy as a possible factor in geographic distribution of *Histoplasma capsulatum*. *Canadian Journal of Microbiology* 13, 1-7.

Stotzky, G. and Rem, L.T. (1967). Influence of clay minerals on microorganisms. IV. Montmorillonite and kaolinite on fungi. *Canadian Journal of Microbiology* 13, 1535-1550.

Sutherland, I.W. (1972). Bacterial exopolysaccharides. *Advances in Microbial Physiology* 8, 143-213.

Tyagny-Ryadno, M.G. (1968). Distribution of microorganisms and nutrients in soil aggregates. *Visnyk Sil's'kogospodar'skoi Navki* 11, 46-51.

Vaughan, D. and Macdonald, I.R. (1976). Some effects of humic acid on cation uptake by parenchyma tissue. *Soil Biology and Biochemistry* 8, 415-421.

Verma, L. and Martin, J.P. (1976). Decomposition of algal cells and components and their stabilization through complexing with model humic acid-type phenolic polymers. *Soil Biology and Biochemistry* 8, 85-90.

Verma, L., Martin, J.P. and Haider, K. (1975). Decomposition of carbon-14-labelled proteins, peptides and amino acids; free and complexed with humic polymers. *Soil Science Society of America Proceedings* 39, 279-284.

Weber, D.F. and Gainey, P.L. (1962). Relative sensitivity of nitrifying organisms to hydrogenious in soils and solutions. *Soil Science* 94, 138-145.

Weber, J.B. (1970). Mechanisms of adsorption of *s*-triazines by clay colloids and factors affecting plant availability. *Residue Reviews* 32, 93-128.

Weber, J.B. (1972). Interaction of organic pesticides with particulate matter in aquatic and soil systems. *Advances in Chemistry Series No. 111. Fate of Organic Pesticides in the Aquatic Environment*. American Chemical Society.

Weber, J.B. and Cobie, H.D. (1968). Microbial decompositon in diquat adsorbed on montmorillonite and kaolinite clays. *Journal of Agriculture and Food Chemistry* 16, 475-478.

Williams, S.T. and Mayfield, C.I. (1971). Studies on the ecology of actinomycetes in soil. III. The behaviour of neutrophilic streptomycetes in acid soil. *Soil Biology and Biochemistry* 3, 197-208.

Wolf, D.C. and Martin, J.P. (1976). Decomposition of fungal mycelia and humic-type polymers containing carbon^{-14} from ring and side-chain labelled 2,4-D and chloropropham. *Soil Science Society of America Journal* 40, 700-704.

THE ACCUMULATION OF ORGANISMS ON THE TEETH

P. RUTTER

Unilever Research, Isleworth Laboratory,
455 London Road, Isleworth, Middlesex TW7 5AB.

INTRODUCTION

The accumulation of organisms on the teeth is a source of major inconvenience to most individuals at least once or twice during their lives. Organisms that collect on and around the teeth are largely responsible for tooth decay and gum disease. Although these disorders are not usually very serious they are widespread throughout the population and can cause considerable discomfort. Eventually these disorders may lead to the loss of teeth or to the requirement for extensive treatment. One approach to the effective control of these diseases lies in understanding the mechanisms that enable organisms to collect in such large numbers on teeth.

The mouth can be thought of as a small vessel with a drainage hole at the base and containing two sets of opposing baffles arranged in semi-circles to enclose a large stirring device. The stirrer is designed to distribute a liquid lubricant, injected through ports in the sides and base of the vessel over the baffles and its inner surfaces. In mechanical terms this is a complex system and it is further complicated by variations in the rate of addition of lubricant and the motion of the stirrer. In biological terms the mouth can be thought of as a warm ecological system in which a number of different types of bacteria thrive on a number of different surfaces, which are periodically bathed in a glycoprotein-rich fluid known as saliva.

SALIVA

Saliva contains a variety of different materials such as proteins, inorganic ions, vitamins [Kauffman *et al.*, 1953], amino-acids [Dreyfus *et al.*, 1968] and lipids [Prout, 1976] dispersed or dissolved in water. The composition of saliva varies between individuals and depends markedly on the time of day and rate at which it is collected. There are also differences in composition between samples taken from the parotid or submaxillary glands, the secretions of which combine to form saliva [Gron, 1973]. The protein concentration of unstimulated parotid saliva has been determined as 232 mg % but this falls dramatically on stimulation by chewing [Dawes, 1972]. In a recent study of the state of calcium and inorganic phosphate in human saliva, Gron [1973] calculated the mean ionic strength of whole saliva to be 45.1 mM, based on the presence of Ca^{2+}, Na^{+}, K^{+}, H^{+}, $CaHCO_3^{+}$, $CaH_2PO_4^{+}$ and the anions HPO_4^{2-} $H_2PO_4^{-}$, OH^{-}, HCO_3^{-} and Cl^{-}. The physical properties of saliva are also variable and may be of great importance in determining the way in which organisms are moved round the mouth.

Saliva in the mouth is a viscous, almost stringy, clear liquid with a tendency to foam and collect at points of contact between the teeth or where the cheeks and gums touch. When the lips or cheeks are drawn away from the teeth a saliva film can sometimes be seen retreating from the flat surfaces of the teeth to the interdental spaces and the gum margins. Various determinations of the physical properties of saliva such as surface tension (53.1 $mN.m^{-1}$ - Glantz, 1970), osmotic pressure, and viscosity [Afonsky, 1961] have been made, although their relevance to such a heterogenous mixture is difficult to ascertain. This difficulty is increased by the gradual loss of glycoprotein carbohydrate due to bacterial enzyme action [Leach and Critchley, 1966] which could result in a loss in solubility of the protein fraction.

In addition to lubricating the mouth, and acting as a transport fluid for oral organisms, saliva appears to play an important role in protecting the tooth enamel. If a tooth is cleaned by abrasion to expose the enamel surface it absorbs a layer of proteinaceous material from the saliva within two seconds [Baier, 1976]. This process continues with time and can eventually lead to the formation of a layer up to 10 μm thick [Meckel, 1965]. This protective layer is known as the enamel pellicle and its rapid formation on any freshly exposed enamel usually ensures that no bacteria ever come into contact with the enamel surface. The tooth surface therefore may be defined (for the purposes of a discussion on bacterial accumulation) as a layer of proteinaceous material [Mayall, 1977] which is in some state of equilibrium with the continuously secreted and changing fluid from which it originated. It is the deposition and accumulation of organisms on this surface that results in the eventual formation of the macroscopic bacterial deposits known as dental plaque.

THE DEVELOPMENT OF PLAQUE

The development of plaque can be regarded as a two-stage process involving the transport of organisms to the tooth (or more correctly the pellicle surface) and the behaviour of the organisms on arrival. Large numbers of organisms are found in saliva. Some organisms may enter the mouth with the diet or from other contact with the environment. However the quantitative contribution of organisms from these sources to the plaque is likely to be small, since plaque will form in tube-fed animals where no food enters the oral cavity [Egelberg, 1965].

The concentration of organisms in saliva depends upon the type of organism and probably the degree of abrasion received by the oral surfaces before sampling. Values of 10^8 colony-forming units (these may be single organisms or small aggregates) to 10^3 colony-forming units per ml of unstimulated saliva, have been determined for different organisms [van Houte *et al.*, 1974]. Simple experiments in our own laboratory have shown that the number of organisms in saliva can increase dramatically after chewing (see Table 1). This is presumably due to the removal of organisms attached to the oral surfaces during the inevitable abrasion which these surfaces receive during chewing.

The total concentration of cells in saliva, therefore, probably fluctuates throughout the day depending on the amount of activity in the mouth. Some organisms might also be able to multiply in saliva [Cowman, 1977; Molan, 1971]. Others, perhaps, can utilise suitable nutrients from the host's diet,

Table 1.

The increase in the concentration of organisms in saliva caused by chewing unflavoured gum

Saliva from volunteer	Total number of organisms in saliva organisms per ml.	
	before chewing	after chewing
1	6.5×10^7	8.5×10^7
2	3.7×10^7	14.0×10^7
3	11.0×10^7	14.0×10^7
4	5.0×10^7	13.0×10^7
5	3.5×10^7	7.0×10^7
6	3.8×10^7	7.2×10^7

which might be maintained at low concentration in saliva due to adsorption followed by gradual desorption from the oral surfaces.

In addition to being influenced by the concentration of organisms in saliva, the rate of deposition of bacteria onto the teeth is almost certainly determined to a large extent by the position on the tooth surface at which contact is first made. Plaque will form on most of the exposed surfaces of the teeth and, more importantly from the point of view of gum disease, along the gum margin.

PLAQUE FORMATION ON THE PLANE SURFACES OF THE TEETH

The simplest sites to consider are the centre-plane surfaces of the front teeth. Organisms will be carried to this position by the saliva which will then drain away to the extremities of the tooth surface. If the thickness of the salivary film immediately prior to draining starts at about 50 μm, the removal or drag forces exerted on particles within the liquid might be expected to increase as the square of the particle diameter, assuming streamline flow conditions [Coulson and Richardson, 1967]. Other hydrodynamic forces might tend to lift large particles off the pellicle surface [Lips and Jessup, this volume]. Large aggregates on the point of depositing onto the pellicle surface would therefore tend to be dragged to the extremities of the teeth with the draining saliva, whereas smaller aggregates or single bacteria might withstand the fluid removal forces and adhere to the pellicle.

Large aggregates would also be more subject to removal by movement from the tongue and cheeks. This movement could also replenish the saliva film. In the absence of replenishment the saliva film thickness will rapidly decrease with time [Uno and Tanaka, 1972] until it forms a liquid/air meniscus moving across the pellicle surface. Any particles in close proximity to the pellicle could be pressed against it as the liquid meniscus passes over them or be swept away if their interaction with the pellicle is weak. Thus bacteria and aggregates not removed by the drag forces within the draining saliva film will eventually remain adhering to the pellicle. Whether or not the organisms remain attached to the pellicle after the saliva meniscus has passed over them depends on the magnitude of any adhesive interaction between the bacterial and pellicle surfaces and possibly the relative humidity of the surroundings.

Bacterial adhesion has been the subject of many papers although most have been restricted to solid/liquid systems rather than the fluctuating solid/liquid, solid/air and solid/air/liquid systems encountered in the mouth. Marshal, Stout and

Mitchell [1971] have suggested that marine organisms are able to attach to surfaces using a two-stage process. The first, or reversible, stage involves the location of the cell near the surface in a weak DLVO secondary minimum (see page 13) produced by a balance of the van der Waals attractive forces and electrostatic repulsion forces. The second, or irreversible, stage is achieved by the cell synthesising polymeric material which is able to adsorb to the surface, thus anchoring the cell and substrate surfaces together.

The studies of Rutter and Abbot [1977] with oral streptococci and of Fletcher [1973] with a marine pseudomonad, suggest that extracellular polymers associated with the cell wall are used initially to anchor these bacteria to the substrate surface by polymer bridging. This suggests that organisms which are capable of interacting with the pellicle surface to form polymer bridges have an advantage over organisms that have to rely on weak van der Waals interactions.

It is known that many organisms are aggregated by saliva and it is suggested that some of the components of the pellicle will aggregate organisms [Magnusson and Ericson, 1973; Hay *et al.*, 1971]. Such components might form bridges between the pellicle and organisms and lead to attachment. Organisms aggregated by saliva, however, might also be more easily removed from the tooth surface by the draining saliva film. The wettability, or hydrophilic/hydrophobic balance, of the cell surface has also been suggested as a method of selectively adsorbing bacteria to surfaces [Marshall, 1976]. This could favour the deposition of certain oral organisms whilst others might be more stable in saliva.

According to the arguments outlined so far, after thorough cleaning the tooth surface would rapidly be covered by a layer of small aggregates or single bacteria that are able to interact with the pellicle surface. The rate of accumulation of bacteria on the pellicle surface would depend on the concentration of organisms in the saliva, their aggregate size distribution and the rate of flow of saliva over the tooth surface. Since at the least the concentration of organisms in saliva is variable, the rate of accumulation of organisms on the tooth surface would also be variable. As soon as the pellicle surface is partially covered with organisms, however, two more factors become important; these are the interaction of organisms in the saliva with those already on the pellicle surface and secondly, the bacterial division rate.

It is known that bacteria will interact with similar and dissimilar organisms to form aggregates [Gibbons and Nygaard, 1970] and artificial plaques [Miller and Kleinman, 1974]. The deposition rate of organisms on to the tooth surface is therefore not only determined by the interaction between organisms and pellicle but also by the interaction between organisms themselves.

The second important factor is bacterial growth. Various figures have been suggested for the growth rates of organisms in the mouth, ranging from one division every two days [Tanzer *et. al.*, 1969] to one every three hours [Socransky *et al.*, 1977]. Under conditions of constant growth rate, the number of bacteria, X, which have originated from X_0 after time t, is easily calculated from the equation

$$X = X_0 e^{\mu t}$$

where μ is the specific growth rate.

Even if the organisms in the mouth were dividing only once every day, this exponential relationship would ensure a one-thousand-fold increase in cells during a period of ten days. This rate of growth coupled with the continuing deposition of organisms would rapidly cause the teeth to be covered and the mouth eventually filled with microbes. The fact that this seldom, if ever, occurs implies that some organisms are continually being removed from the teeth and oral surfaces and swallowed. Their removal from the oral surfaces has already been suggested as being responsible for the increase in the number of organisms in saliva after chewing.

The removal of organisms could operate in one of two ways, either at a constant ratee or at a variable one. If it occurs at a constant rate the plaque would simply take longer to fill the mouth since the balance of variable deposition against constant removal would either result in no plaque at all or a gradually increasing level of plaque. On the other hand, if the removal rate was variable or removal became more effective as the plaque got thicker and possibly less cohesive, plaque would build up to a level at which deposition and removal were in equilibrium. Growth might then be the controlling factor until this too came into equilibrium with removal.

In this latter stage of plaque formation, however, growth might favour some organisms at the expense of others and lead to an alteration in the structure of plaque. This process would inevitably result in some degree of plaque structuring such as that shown by Listgarten [1976]. The cohesive strength of the structured regions might exceed that of the randomly deposited cells, consequently the structured plaque would be more stable. As such a plaque began to extend farther away from the tooth surface however, a point would probably be reached where even this cohesion is overcome and some organisms are removed.

The accumulation of cells on the teeth is therefore almost certain to be a complex dynamic process resulting in an eventual equilibrium in which the distribution, deposition and growth rate of organisms are in fluctuating balance with the removal rate.

THE ACCUMULATION OF SUPRAGINGIVAL PLAQUE IN VIVO

A study by Socransky *et al.* [1977] of the developing supragingival plaque has shown that the number of organisms on the middle third of the plane outer (buccal) surface of the upper right first molar tends to a maximum of 10^8 organisms cm^{-2} after two to four days, although the composition of the deposit continues to change for some time after this period. Samples after five and fifteen minutes, revealed 5×10^5 organisms cm^{-2}, a value which remained constant for about eight hours. There was then a rapid increase in number to a value of 10^8 organisms cm^{-2} after forty-eight hours.

Socransky explained these results in terms of an initial colonisation phase resulting in 0.1 to 1.0% coverage of the tooth surface taking place between nought and eight hours, followed by a rapid growth phase resulting in a total coverage of 10^8 organisms cm^{-2}, and finally a remodelling phase during which the composition of the plaque changes. An alternative explanation would be to assume that after cleaning, very rapid deposition results in the accumulation of 5×10^5 organisms cm^{-2} of pellicle surface within five minutes. This is in equilibrium with a removal process which maintains the coverage at this level for eight hours. Overnight the removal of organisms might be reduced owing to low mouth activity. If the deposition rate were maintained, however, the coverage after eight hours sleep would be 480×10^5 organisms cm^{-2}. The accumulated organisms might also have grown to some extent and consolidated into a plaque that is able to maintain this degree of coverage during the next period of mouth activity.

After twenty-four hours the processes of deposition growth and removal would continue, but probably at different rates since the pellicle surface is already partially covered with organisms and consequently interbacterial interactions become more important. If this leads to an increase in the bacterial/surface collision efficiency it might also help to establish the new and apparently stable equilibrium coverage of 10^8 organisms cm^{-2} after forty-eight hours. In order to achieve an increase from 10^6 organisms cm^{-2} after twenty-four hours to 10^8 organisms cm^{-2} after forty-eight hours by growth alone, the bacteria would need to double every 3.6 hours. When growth, deposition and removal finally come into equilibrium, the plaque may become structured [Listgarten, 1976] by the growth of certain organisms.

A study by Lie [1977] using hydroxyapatite and epoxy resin composite splints inserted into the mouth to cover the buccal surfaces of maxillary molars, showed large numbers of individual organisms distributed over the surface after six hours. After forty-eight hours the coverage was much greater and was suggested as being due to growth. The cells, however, appeared to be well separated and showed patterns similar to the deposition of polystyrene latex spheres or streptococci on a polystyrene surface (see Fig.1). The apparent colonies shown by Lie might therefore be groups of deposited cells collected on energetically favourable areas of the splint surface.

Thus the accumulation of organisms on the planar regions of the teeth can be regarded as the result of a balance of deposition, growth and removal. The relative importance of each process may vary throughout the day but eventually, usually within two or three days, an equilibrium coverage of the teeth is achieved. This final coverage will probably vary from individual to individual and depend upon the plaque cohesion, the growth rate of the plaque organisms, the activity of the mouth, the composition of the saliva (which may affect bacterial adhesion) and the physical properties of the saliva. Perhaps even the relative humidity at the tooth surface may play a role on the labial surfaces of the front teeth.

THE ACCUMULATION OF BACTERIA AT THE SOLID/LIQUID/AIR INTERFACE ALONG THE GINGIVAL MARGIN

Perhaps the most common site in the mouth at which the formation of plaque has been studied is the line along which the gums meet the teeth (the gingival margin). The presence of organisms in this region is thought to be responsible for the mild gum inflammation which is often indicated by slight bleeding when the teeth are well brushed. This can lead to more serious diseases if not checked. Two factors distinguish the environment of the gum margin from the buccal surfaces of the teeth. One is the actual shape of the gum margin and the other is the presence of gingival fluid which passes into the saliva from the bottom of the gingival crevice.

Fig. 2 is a diagram of a cross-section through a tooth and the surrounding gum tissue. The striking difference between this and other regions of the tooth surface is the presence of a 'pool' of saliva retained in the gingival crevice by its surface tension. This could dramatically alter the process of deposition in this region since there are two points at which a solid/liquid/air interface is formed. This is important because the deposition of particles at a solid/liquid/air interface is often completely different from the deposition at the solid/liquid interface in the same system.

If, for example, a clean glass microscope slide is half immersed in a suspension of oral organisms for a few minutes

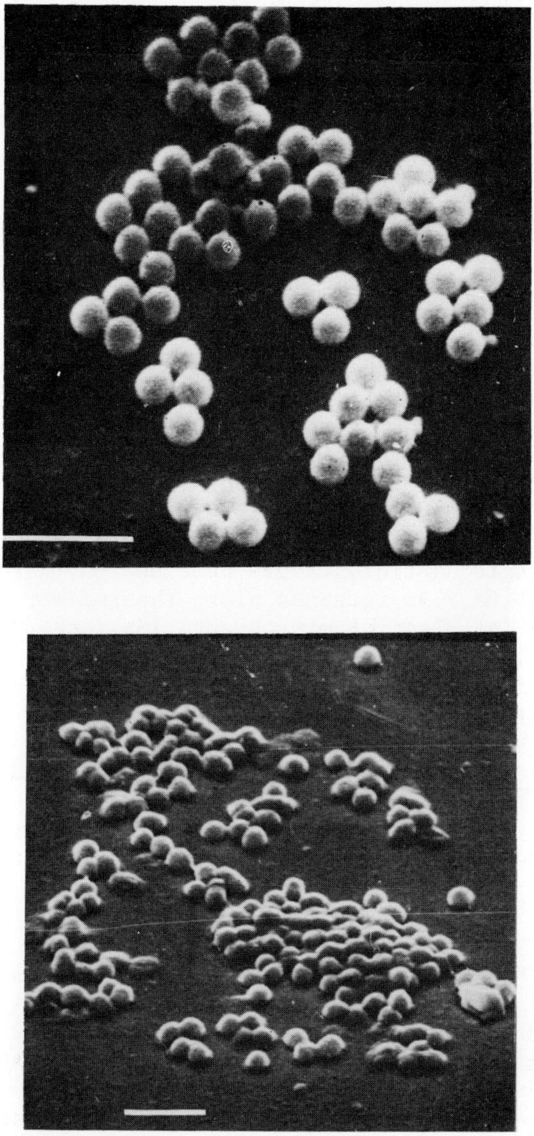

Fig. 1. Scanning electron micrographs to show the deposition patterns of polystyrene latex particles (Fig. 1a, top) and Streptococcus salivarius (Fig. 1b, bottom) on a smooth polystyrene surface. The particles were allowed to deposit for ten minutes from a 0.15M-saline suspension onto the surface of a disc rotating at 180 rpm. (Bar represents 4 microns).

before being removed and stained with crystal violet, a distinct line of cells is seen on the surface of the slide corresponding to the junction of the glass, suspension and air. The rapid formation of a line of deposited organisms in this region may be due to thermal and flotation effects causing the movement of particles into the liquid meniscus. The amount of deposition in this region is apparently determined by the concentration of organisms in the suspension, the rate of flow of suspension past the solid surface, the contact angles of the continuous medium on the solid surface (θ_s) and bacterial surface (θ_b), and the contact time [Rutter and Leech, 1977].

One of the most striking features of deposition at the solid/liquid/air interface is the speed at which a deposition line is formed, especially if the concentration of organisms in suspension is high (of the order of 10^8 organisms ml^{-1}). The other feature of deposition in this region is the inclusion of aggregates in the deposition line together with single particles. Thus, if solid/liquid/air deposition were of relevance in the oral situation, the rapid deposition of aggregates and single particles would be expected in the region where the saliva film retained along the gum margin meets the teeth. Scanning electron micrographs [Saxton, 1975] of bacterial globules, which appear on the tooth surface close to the gum margin a few minutes after thorough cleaning could be evidence of this type of deposition process.

The apparent dependence of deposition at the solid/liquid/air interface on contact angle can also apply a degree of selectivity to the type of organism that is deposited. Earlier in this chapter it was suggested that the ability to adsorb pellicle components to the cell surface would enable certain organisms to colonise the buccal surfaces of the teeth. In the gingival margin, however, the simple physical wettability of the cell surface could determine the ease and speed of adsorption. This is not a new concept but has been implicated in bacterial distribution and phagocytosis since 1924 [Mudd and Mudd, 1925; van Oss and Gillman, 1972].

If a small particle, with a hydrophobic surface immersed in an aqueous solution, approaches the liquid/air interface with sufficient energy to break through the surface molecules of water, the particle will be trapped in the liquid meniscus. This is the principle of flotation and relies on the balance of the downward pull of the particle weight by the upward pull of the liquid meniscus. In the case of large values of contact angle θ, relatively large objects like waxed sewing needles can be floated on water.

Owing to the small size of bacteria, a very small contact angle is sufficient to float the cell in the liquid meniscus.

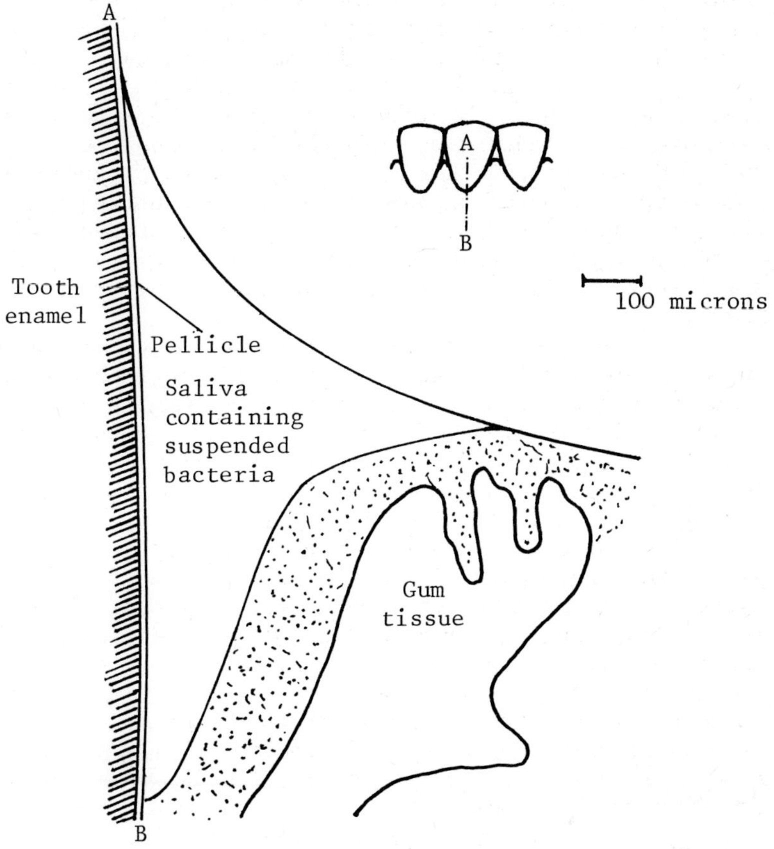

Fig. 2. Diagram of a cross-section (see inset) through the gum margin showing a healthy 0.8 mm deep crevice containing saliva.

Therefore, bacteria with a finite contact angle probably collect in the liquid meniscus whereas bacteria that are completely wetted (that is where $\theta_b = 0$) do not. When a solid surface comes into contact with a liquid meniscus containing trapped particles it rapidly becomes contaminated at the solid/liquid/air junction owing to the high concentration of particles in this region. Furthermore, it would be most rapidly colonised by the bacteria with finite contact angles.

Results of experiments carried out with *Streptococcus salivarius* and *Streptococcus mitior* tend to support this selectivity of adsorption, showing that the deposition at the glass/

saliva/air interface is strikingly different (Fig.3). Both these organisms are found in saliva but *Streptococcus mitior* is found in considerably higher numbers in dental plaque [Gibbons and van Houte, 1973].

Saxton [1975] has strongly implicated the presence of gingival exudate in the gingival margin as an important factor in the rapid build up of gingival plaque. In fact the rate of formation of globules on the tooth surface appears to be inversely related to the health of the gingiva and hence the quantity of gingival exudate. This implicates the crevicular fluid, or a component of the fluid, in the deposition of organisms in this region. Crevicular fluid may be involved in two ways: by providing suitable bridging molecules for the organisms to anchor themselves to the tooth surface as described previously, or by providing materials that adsorb to the bacterial surfaces and increase the contact angles. This latter phenomenon is often used in flotation in which mineral particles are separated from mixtures by conditioning their surfaces with surfactants to the correct contact angle [Davies and Rideal, 1963].

Another factor which might encourage the accumulation of organisms in the region of the gingival margin is the relatively protected location of this site. Organisms collecting in this region might not be subjected to the abrasive and liquid drag forces experienced by organisms attached to the more exposed areas of the tooth surface. In fact organisms removed from the exposed tooth surfaces may be dragged into the gingival regions by draining saliva films, thus maintaining a high concentration of organisms in the saliva retained in this region.

The mechanical abrasion of the tooth surface during eating and chewing might also tend to sweep organisms along the tooth surfaces and into the gum region. Surveys attempting to correlate gingival health with the consumption of apples and carrots [Birkeland and Jorkjend, 1974] have shown that no benefit can be gained from eating these foodstuffs. However, due to the limitations of the methods used to assess this correlation it is impossible to say at this stage whether organisms are removed from the teeth or impacted in the gingival region during the chewing of hard or fibrous foods.

SELECTION

Any organism present in saliva, providing it is not swallowed, may eventually come into contact with a surface. Its interaction with that surface can either lead to attraction or repulsion. If it leads to the former the attraction can either be strong (involving several different types of chemical or physical bonding) or weak (perhaps only involving dispersion

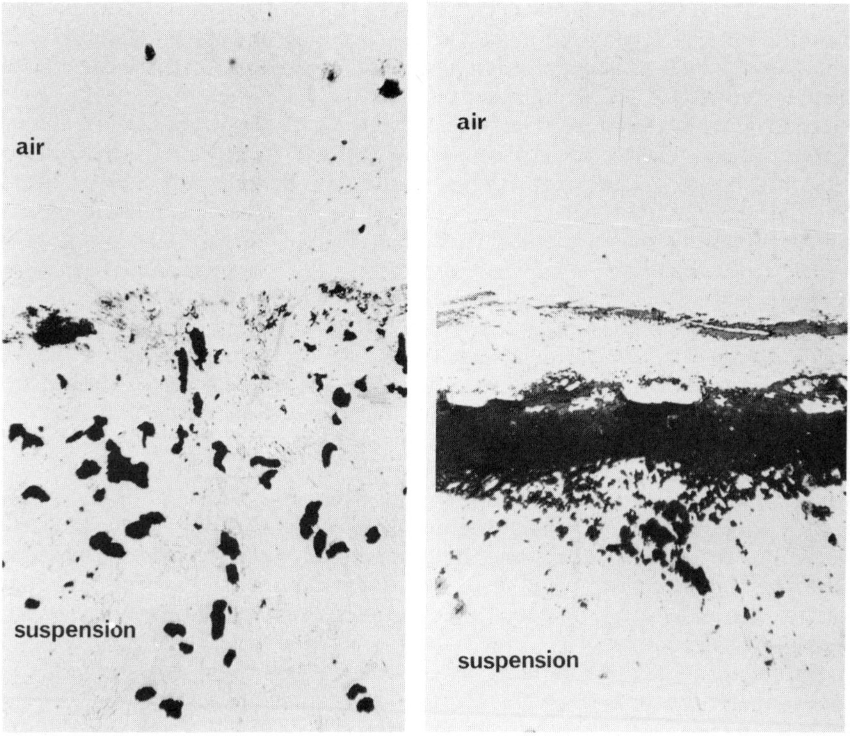

Fig. 3. Photographs to show the accumulation of (left) Streptococcus salivarius, (right) Streptococcus mitior, at the glass/saliva/air interface on a microscope slide partially immersed in a saliva suspension of streptococci (10^7 organisms per ml) for ten minutes.

forces). In the presence of removal forces the weakly bound organisms will be removed and eventually be replaced by bacteria capable of strong interaction. It is possible that all the oral bacteria are weakly attracted to the surfaces of the mouth, which might account for the occasional presence of *Streptococcus salivarius* in the plaque or *Streptococcus mutans* on the tongue [Gibbons and van Houte, 1973].

Most oral bacteria however, appear to be capable of strong interaction with certain specific surfaces although the nature of this strong interaction is unclear. Parsegian and Gingell [1972] have shown that it is possible to explain the selective adhesion between two model cell membranes with 'fuzzy' coats using the theories of colloid stability. Ninham and Richmond [1973] have gone further and calculated the conditions under which a close substrate, foreign body or neighbouring cell will cause dissolution or build-up of an extra-cellular layer or the formation of a tight junction or cell glue. These interactions can be quite specific and it appears that it may be unnecessary to invoke additional mechanisms such as calcium or polymer bridging to account for cell adhesion [Ninham and Richmond, 1973]. Polymer bridging however may play an important contribution in the deposition of bacteria especially if the extracellular polymers are hydrated and extend some distance into the suspending medium. In this case a situation analogous to the polymer flocculation described by Fleer and Lyklema [1974] may occur.

One way of approaching the problem of the involvement of extracellular polymers in bacterial adhesion is to consider bacteria as particles which are coated, to a lesser or greater extent, with 'fuzzy' polysaccharide layers [Dudman, 1977]. The conformation, composition, solubility and molecular weight of the extracellular polysaccharides may all vary. These sources of variation determine their ability to adsorb to inert surfaces [Freedman and Tanzer, 1974] and to interact specificially or non-specifically with complimentary polysaccharides or proteins [Thom *et al.*, 1977; Edwards 1978].

The situation is complicated further, in some cases, by the presence of free polysaccharide released into the environment [Corpe *et al.*, 1976] and the presence of enzymes on the bacterial cell surface which are capable of synthesising and adsorbing or binding polysaccharides [Gibbons and Fitzgerald, 1969]. In the oral cavity a considerable amount of work has been directed towards relating cell surface polysaccharides to the adhesion of *Streptococcus mutans*. The complexity of the system and the methods used to evaluate adhesion has led to some confusion concerning the relationships between cell surface enzymes and polysaccharides.

A number of authors have demonstrated the importance of

insoluble extracellular dextran, synthesised from sucrose, in the irreversible attachment of *Streptococcus mutans* to hard surfaces [Schachtele *et al.*, 1975; Mukasa and Slade, 1974; Gibbons and van Houte, 1973; Olson *et al.*, 1972]. Mukasa and Slade [1973] suggested that adhesion required the synthesis of an insoluble dextran-levan polymer by a cell bound enzyme and the binding of the polymer between a polymer receptor site, probably protein [Kelstrup and Funder-Nielson, 1974], and the substrate. Synthesis of the polymer from sucrose in the presence of the cells was required for adherence which indicated that an 'active' form of the polymer was required. Polymer synthesised by cell-free *Streptococcus mutans* enzymes did not cause adherence when added to *Streptococcus mutans* cells. More recent work with mutants has shown that the adhesive ability of this organism is directly related to the type of insoluble extracellular polysaccharide produced [Johnson, 1977].

There is evidence however, that aggregation (or intercellular adhesion) does not require the synthesis of dextran, merely its presence [Kelstrup and Funder-Nielsen, 1974]. Kuram itsu [1975] has also shown that active cell bound dextran sucrase (enzyme) is not required for cell binding to pre-formed dextran layers. This apparent difference in behaviour led Freedman *et al.* [1974] and Nalbandian *et al.* [1974] to suggest that aggregation and adhesion were separate and dissociable traits.

McCabe [1976, 1977] has summarised the situation in a similar way to Olson [1974] by suggesting that the adhesion and aggregation of *Streptococcus mutans* has only two requirements. First that adherent dextrans be formed by enzyme molecules which are in close proximity to the surface, either as components on a cell momentarily adsorbed to the surface by a non-specific mechanism or as cell free molecules which have adsorbed to the surface, and second that the cells specifically bind to these adherent dextrans via a cell-surface dextran receptor. McCabe [1977] stated further than in both aggregation and cell to surface adhesion dextransucrase primarily serves to synthesise dextran while binding of dextran occurs at a unique cell surface site. This does not confirm the generally held view that polysaccharide is necessary for the adhesion of *Streptococcus mutans* however, since Ellwood *et al.* [1974] have shown that the adhesiveness of glucose grown *Streptococcus mutans* can be related to growth rate. Only small amounts of polysaccharide were produced under these conditions.

In the oral cavity this already complex picture is further complicated by the presence of dextranase producing bacteria in dental plaque and saliva which have the capacity to modify the colonisation of the tooth surface by *Streptococcus mutans* [Schachtele, 1977]. Antibodies [Evans *et al.*, 1977] may also interfere with the adhesion of *Streptococcus mutans* by inhibiting dextran synthesis [Genco and Evans, 1974], blocking the cell surface receptors specific for the transferases or for the enzyme-dextran complexes or thirdly binding to the surface antigens which in turn alter the bacterial surface properties.

The fact that saliva and the surface which the organisms eventually colonise, directly influence bacterial adhesion has been recognised for some time. Saliva, for example, contains aggregation inducing substances (AIS) which will aggregate *Streptococcus sanguis* and *Streptococcus mitior* when added to buffered suspensions of the organism [Kashket and Donaldson, 1972; Hay, Gibbons and Spinnel, 1971]. Saliva will also aggregate *Streptococcus mutans* [Magnusson and Ericson, 1973] under similar conditions. Aggregation inducing substances responsible for bacterial aggregation can be adsorbed quantitatively to hydroxyapatite (HA) [Magnusson and Ericson, 1973,1976] and can therefore be expected to adsorb to enamel but not necessarily to pellicle.

When saliva is adsorbed to HA it inhibits the adhesion of *Streptococcus mutans* [McGaughey *et al.*, 1971; Ericson, 1975]. When adsorbed to enamel however, saliva can enhance bacterial adsorption [Hillman, 1970; Ørstavick, 1974]. This observation is further complicated by results obtained by Sonju [1977] who showed that differences in the amino acid composition of pellicles formed on enamel, composite and amalgam are reflected by differences in the number and type of organisms which colonise them.

If saliva itself is used as a suspending medium for the bacteria as it is in the mouth the adhesion of bacteria to enamel tends to be reduced [Ørstavick, 1974]. Also, if certain strains of streptococci are inoculated with salivary glycoprotein their adhesion to epithelial cells is reduced [Williams, 1975]. Work carried out in the absence of saliva suggests that the adhesion of bacteria to the oral mucosa involves the extracellular 'fuzzy' coat [Liljemark and Gibbons, 1972]. Adhesion to oral epithelial cells can be blocked by concanavalin A or serum IgA [Williams and Gibbons, 1975]. Concanavalin A also appears to enhance the saliva mediated aggregation of oral organisms and also to enhance the adsorption of organisms by saliva coated HA [Kashket, 1975].

The studies briefly reviewed in this section demonstrate the difficulty involved in producing a simple model capable of describing the effect of extracellular polysaccharides on selective adhesion. A number of adhesive interactions have been demonstrated in isolation as, for example, the insoluble polysaccharide production by *Streptococcus mutans*. However, when the system is studied in the presence of saliva, other bacteria, antibodies and pellicle, the relevance of the polysaccharide production becomes less certain. In such a complex system the contribution of each component must be evaluated and judged relative to the others.

The final adhesive interaction is probably a combination of several mechanisms influenced possibly by the overall reaction of the host. The level of calcium in the host saliva, for example, may affect a number of potentially adhesive interactions producing an apparent overall calcium dependence for adhesion [Rolla, 1977]. If a general model is required then it must be very simple and consist possibly of particles surrounded by polymer layers suspended in a fluid containing more polymers. These particles would then colonise surfaces covered in more adsorbed polymers. Adhesion can then be regarded as the specific or non-specific interaction of these polymers in such a way as to anchor the bacteria to each other or to a surface. Adhesion by this mechanism has the advantage of being able to operate at separations greater than those required to invoke electrostatic repulsion providing enough polymer contacts are made to overcome the removal forces continually operating in the oral environment.

Selective colonisation could therefore arise because the difference between the weak non-specific interactions between bacteria and surfaces and the strong polymeric interactions is sufficiently large for the removal forces operating in the mouth to continually rearrange the organisms which have deposited.

THE LONG TERM ACCUMULATION OF PLAQUE

The arguments presented above suggest that the accumulation of organisms on the exposed planar surfaces of the teeth and along the gingival margins are both important. Both sites accumulate organisms rapidly although large numbers of deposited organisms might occur at the solid/liquid/air interface more readily than at the solid/liquid interface. This is due to the absence of removal forces and the deposition of aggregates as well as to the apparent speed of deposition inferred from experiments carried out *in vitro*.

In the absence of any oral hygiene the teeth eventually become almost completely coated with a waxy-looking film of organisms. Under these conditions the accumulated organisms may be dried onto the exposed surfaces of the teeth in the same way that organic deposits dry onto plates and cups. This could lead to a reversal in the favourability of the two accumulation sites. Along the gingival margin the organisms might be more hydrated and hence more easily removed than on the buccal surfaces of the teeth where drying can occur.

An experiment carried out by Harrap [1976] suggests that

this is in fact the case. Plaque was allowed to accumulate on the buccal surfaces of the two front incisors in the absence of brushing for a period of eight days. The plaque was then stained with a red dye and the teeth brushed. Photographs taken at increasing time intervals during the brushing show that the most difficult plaque to remove was situated in the exposed central region of the tooth (Fig.4). Thus although organisms may be removed from the buccal surfaces during the initial stages of plaque formation when the deposit is small compared with the thickness of the salivary film coating the teeth, their removal is more difficult when an appreciable number of cells has accumulated and formed a cohesive layer by drying.

Plaque will accumulate on most of the tooth surfaces and perhaps the accumulation at the gingival margin and on the buccal surfaces is not the most significant in terms of the pathological effects of plaques. For example, as far as the diseases associated with plaque are concerned perhaps the most significant accumulations occur inside the gingival crevice, in fissures in the enamel surfaces and between the teeth. The pathology of these diseases is not well understood and whilst it is possible to say that acid production by plaque bacteria has been strongly implicated in the development of dental decay, the factor or factors responsible for gum disease have yet to be identified. However, the elimination of bacteria from the mouth in the extreme case, or a substantial reduction in plaque quantity, tends to lead to an elimination or reduction in the severity of the diseases. Thus, whilst it cannot be said that the organisms forming the initial deposits of plaque along the gingival margin are responsible for gum disease it is possible to say that in the absence of plaque no disease occurs [Loe, 1969].

CONCLUSION

The formation of bacterial deposits or plaque on the teeth is a problem that is not only restricted to man but is widespread in the animal kingdom [Dent, 1976]. Two aspects of plaque are of interest in terms of the destruction it causes in the oral cavity. The first is the rate of accumulation of plaque organisms on the teeth and the second is tooth and tissue degeneration caused by the presence of plaque and the action of its metabolites.

The rate of accumulation of organisms on the teeth is controlled by a number of parameters. These include the number of organisms present in saliva, their stability when suspended in saliva (closely related to their wettability) and their ability to adhere irreversibly and grow on the pellicle surface.

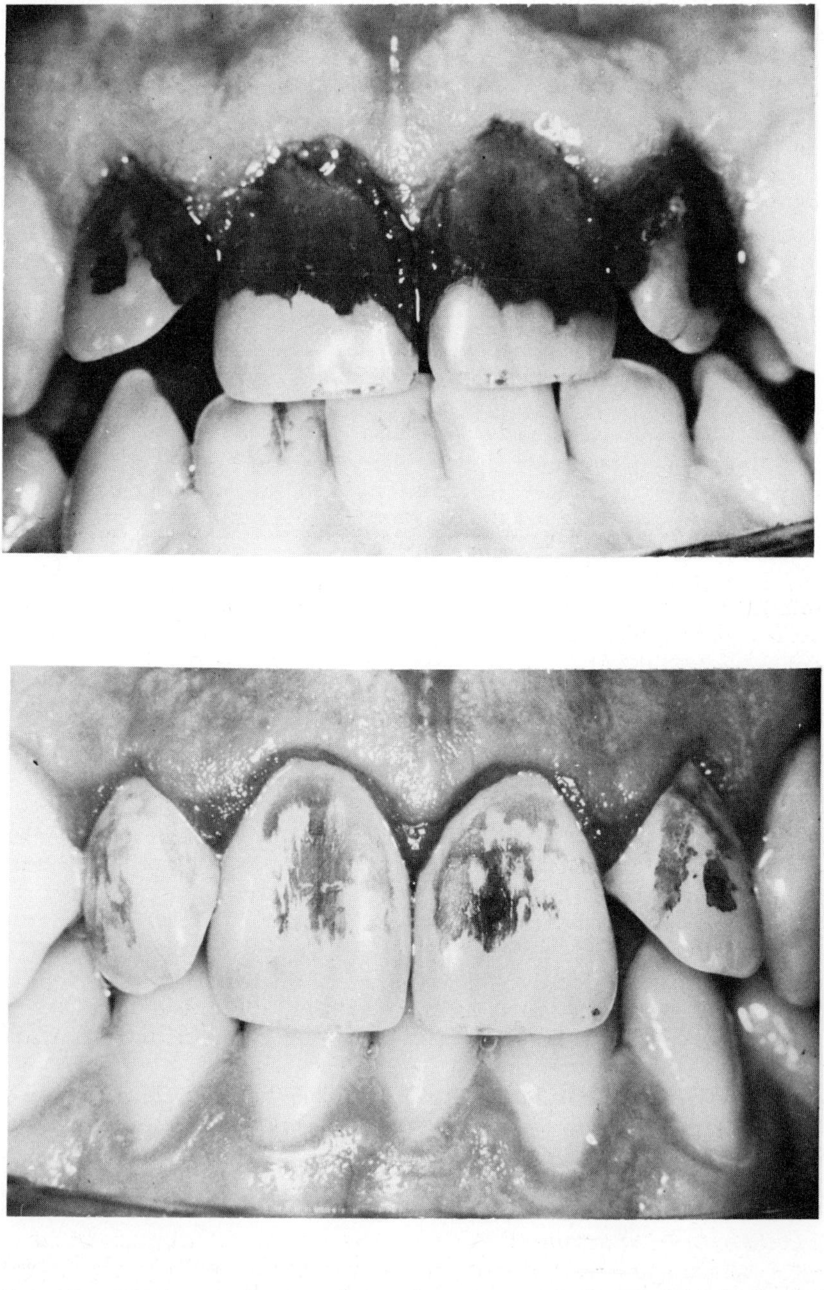

Fig. 4. Photographs to show the removal of established eight day plaque by brushing. Top, before brushing; bottom, after brushing.

The position at which the organisms collect is also of importance because of the high rate of deposition at the solid/liquid/air interface. These parameters are balanced to some extent by a number of randomising factors, the most significant of which are due to the removal forces produced by mastication and saliva flow, but might also include the death of organisms and structural changes affecting the cohesion of the deposit. The cohesion of the plaque is also probably affected by diet since some plaque organisms are able to utilise sucrose in the diet to produce a glue like polysaccharide [Gibbons and van Houte, 1973] which has been strongly implicated in bacterial attachment and coagulation.

The importance of bacterial adhesion in locating organisms on surfaces is clear but the mechanisms involved in colonisation rely not only on the relative strengths of the adhesive interaction and the removal forces but also on the relationship between the organism and the suspending medium. Extrapolation from *in vitro* experiments designed to measure the adhesive interaction alone to the *in vivo* situation should therefore be made with caution.

REFERENCES

Afonsky, D. (1961). *Saliva and its relation to oral health.* Alabama : University of Alabama Press.

Baier, E. and Glantz, P.O. (1976). Characterisation of oral *in vivo* films formed on a variety of different types of solid surfaces. *Journal of Dental Research* 56, Abstract 531.

Birkeland, J.M. and Jorkjend, L. (1974). The effect of chewing apples on dental plaque and food debris. *Community Dentistry and Oral Epidemiology* 2, 161.

Corpe, W.A., Matsuuchi, L. and Armbruster, B. (1976). Secretion of adhesive polymers and attachment of marine bacteria to surfaces. In *Proceedings of the Third International Biodegradation Symposium* pp. 433-442. Edited by J.M. Sharpley and A.M. Kaplan, London : Applied Science Publishers.

Coulson, J.M. and Richardson, J.F. (1967). *Chemical Engineering* vol. II, Seventh Edition. Oxford : Pergamon Press.

Cowman, R.A., Fitzgerald, R.J., Perrella, M.M. and Cornell, A. M. (1977). Human saliva as a nitrogen source for oral streptococci. *Caries Research* 11, 1.

Critchley, P., Saxton, C.A. and Bowen, W.M. (1972). Effect of restricting oral intake to invert sugar or casein on the polysaccharide content of plaque in monkeys *(Macaca irus). International Association of Dentistry Research Abstract 157. Journal of Dental Research* 51, 1284.

Davies, J.T. and Rideal, E.K. (1963). *Interfacial Phenomena.* New York : Academic Press.

Dawes, C. and Chebib, F.S. (1972). The influence of previous stimulation and the day of the week on the concentration of protein and the main electrolytes in human parotid saliva. *Archives of Oral Biology* 17, 1289.

Dent, V.E., Hardie, J.W. and Bowden, G.H. (1976). A preliminary study of dental plaque on animal teeth. *International Association of Dentistry Resarch Abstract 85 Journal of Dental Resarch* 55, Special Issue.

Dreyfus, P.M., Levy, H.L. and Efron, M.L. (1968). Concerning amino acids in saliva. *Experimentia* 24, 447-8.

Dudman, W.F. (1977). Surface polysaccharides in natural environments. In *Surface carbohydrates of the prokaryotic cell*, pp. 358-414. Edited by I. Sutherland. London : Academic Press.

Edwards, P.A.W. (1978). Differential cell adhesion may result from non-specific interactions between cell surface glycoproteins. *Nature, London* 271, 248-249.

Egelberg, J. (1965). Local effects of diet on plaque formation and development of gingivitis in dogs III. Effect of frequency of meals and tube feeding. *Odontologisk Revy* 16, 50.

Ellwood, D.C., Hunter, J.R. and Longyear, V.M.C. (1974). Growth of *Streptococcus mutans* in a chemostat, the ability of the organisms to stick to surfaces, glucose utilisation and acid production at different dilution rates. *Archives of Oral Biology* 19, 659-664.

Ericson, T., Sandham, J. and Magnusson, I. (1975). Sedimentation method for studies of adsorption of microorganisms onto apatite surfaces *in vitro*. *Caries Research* 9, 325-332.

Evans, R.T., Genco, R.J., Emmings, F.G. and Linzer, R. (1977). Antibody in the prevention of adherence; measurement of antibody to purified carbohydrate of *Streptococcus mutans* with enzyme linked immunosorbent assay. In *Microbial Aspects of Dental Caries*, vol. 2, pp. 375-386. Edited by H.M. Stiles, W.J. Loesche and T.C. O'Brien. London : Information Retrieval Ltd.

Fleer, G.J. and Lyklema, J. (1974). Polymer adsorption and its effect on the stability of hydrophobic colloids. *Journal of Colloid and Interface Science* 46, 1-12.

Fletcher, M. and Floodgate, G.D. (1973). An electron-microscope demonstration of an acidic polysaccharide involved in the adhesion of a marine bacterium to solid surfaces. *Journal of General Microbiology* 74, 325-334.

Freedman, M.L. and Tanzer, J.M. (1974). Dissociation of plaque formation from glucan-induced agglutination in mutants of *Streptococcus mutans*. *Infection and Immunity* 10, 189-196.

Genco, R.J. and Evans, R.T. (1974). Specificity of antibodies to *Streptococcus mutans;* significance in inhibition of adherence. The immunoglobulin system. *Advances in Experimental Medicine and Biology* 45, 327-336.

Gibbons, R.J. and Fitzgerald, R.J. (1969). Dextran-induced agglutination of *Streptococcus mutans* and its potential role in the formation of microbial dental plaques. *Journal of Bacteriology* 98, 341-346.

Gibbons, R.J., Kapsimalis, B. and Socransky, S.S. (1964). The source of salivary bacteria. *Archives of Oral Biology* 9, 101-103.

Gibbons, R.J. and Nygaard, M. (1970). Interbacterial aggregation of plaque bacteria. *Archives of Oral Biology* 15, 1397.

Gibbons, R.J. and van Houte, J. (1973). Formation of dental plaques. *Journal of Periodontology* 44, 347-360.

Glantz, P.O. (1970). The surface tension of saliva. *Odontologisk Revy* 21, 119.

Gron, P. (1973). The state of calcium and inorganic orthophosphate in human saliva. *Archives of Oral Biology* 18, 1365-1378.

Harrap, G.J. (1976). Personal communication.

Hay, D.I., Gibbons, R.J. and Spinnell, D.M. (1971). Characteristics of some high molecular weight constituents with bacterial aggregating activity from whole saliva and dental plaque. *Caries Research* 5, 111-123.

Hillman, J.D., van Houte, J. and Gibbons, R.J. (1970). Sorption of bacteria to human enamel powder. *Archives of Oral Biology* 15, 899-903.

Johnson, M.C., Bozzola, J.J., Schelmeister, I.L. and Shklair, I.L. (1977). Biochemical study of the relationship of extracellular glucan to adherence and cariogenicity in *Streptococcus mutans* and an extracellular polysaccharide mutant. *Journal of Bacteriology* 129, 351-357.

Kashket, S. and Donaldson, C.G. (1972). Saliva induced aggregation of oral streptococci. *Journal of Bacteriology* 112, 1127-1133.

Kashket, S. and Guilmette, K.M. (1975). Aggregation of oral streptococci in the presence of Concanavalin A. *Archives of Oral Biology* 20, 375-379.

Kauffman, S.L., Kasai, G.J. and Koser, S.A. (1953). The amounts of folic acid and vitamin B_6 in saliva. *Journal of Dental Research* 32, 840-849.

Kelstrup, J. and Funder-Nielsen, T.D. (1974). Adhesion of dextran to *Streptococcus mutans*. *Journal of General Microbiology* 81, 485-489.

Kuramitsu, H.K. (1974). Adherence of *Streptococcus mutans* to dextran synthesized in the presence of extracellular dextranase. *Infection and Immunity* 9, 764-765.

Kuramitsu, H.K. and Ingersoll, L. (1976). Differential inhibition of *Streptococcus mutans in vitro* adherence by anti-glucosyltransferase antibodies. *Infection and Immunity* 13, 1775-1777.

Leach, S.A. and Critchley, P. (1966). Bacterial degradation of glycoprotein sugars in human saliva. *Nature, London* 209, 506.

Lie, T. (1977). Early dental plaque morphogenesis. A scanning electron microscope study using the hydroxyapatite splint model and a low sucrose diet. *Journal of Periodontal Research* 12, 73-89.

Liljemark, W.F. and Gibbons, R.J. (1972). Proportional distribution and relative adherence of *Streptococcus mitior (mitis)* on various surfaces in the human oral cavity. *Infection and Immunity* 6, 852-859.

Listgarten, M.A. (1976). Structure of the microbial flora associated with periodental health and disease in man. *Journal of Periodontology* 47, 1-19.

Loe, H. (1969). *A Review of the Prevention and Control of Plaque in Dental Plaque.* Edited by W.D. McHugh. Edinburgh and London : E and S. Livingstone.

McCabe, M.M. (1976). Comments on the adherence of *Streptococcus mutans*. *Journal of Dental Research* 55, 226-228.

McCabe, M.M., Haynes, A.V. and Hamelik, R.M. (1977). Cell adherence of *Streptococcus mutans*. In *Microbial Aspects of Dental Caries*, vol. 2, pp. 413-424. Edited by H.M. Stiles, W.J. Loesche and T.C. O'Brien. London : Information Retrieval Ltd.

McGaughey, C., Field, B.D. and Stowell, E.C. (1971). Effects of salivary proteins on the adsorption of cariogenic streptococci to hydroxyapatite. *Journal of Dental Research* 50, 917-922.

Magnusson, I. and Ericson, T. (1973). Affinity for hydroxyapatite of soluble salivary substances causing bacterial aggregation. *Journal of Dental Research* 52, Special Issue 192, Abstract 533.

Magnusson, I. and Ericson, T. (1976). Effect of salivary agglutins on reactions between hydroxyapatite and a serotype C strain of *Streptococcus mutans*. *Caries Research* 10, 273-286.

Marshall, K.C. (1976). *Interfaces in Microbial Ecology*. Cambridge, Mass. : Harvard University Press.

Marshall, K.C., Stout, R. and Mitchell, R. (1971). Mechanisms of the initial events in the sorption of marine bacteria to surfaces. *Journal of General Microbiology* 68, 337-348.

Mayall, C.W. (1977). Amino acid compositions of experimental salivary pellicles. *Journal of Periodontology* 48, 78-92.

Meckel, A. (1965). The formation and properties of organic films on teeth. *Archives of Oral Biology* 10, 585.

Miller, C.H. and Kleinman, J.L. (1974). The effect of microbial interactions on the *in vitro* plaque formation by *S. mutans*. *Journal of Dental Research* 53, 427-433.

Molan, P.C. and Hartles, R.L. (1971). The nature of the intrinsic salivary substrates used by human oral flora. *Archives of Oral Biology* 16, 1449.

Mukasa, H. and Slade, H.D. (1973). Mechanism of adherence of *Streptococcus mutans* to smooth surfaces 1. Roles of insoluble dextran levan synthetase enzymes and cell wall polysaccharide antigen in plaque formation. *Infection and Immunity* 8, 555-562.

Mukasa, H. and Slade, H.D. (1974). Mechanism of adherence of *Streptococcus mutans* to smooth surfaces 2. Nature of the binding site and the adsorption of dextran levan synthetase enzymes on the cell wall surface of the streptococcus. *Infection and Immunity* 9, 419-429.

Mudd, S. and Mudd, E.B.H. (1924). Certain interfacial tension relations and the behaviour of bacteria in films. *Journal of Experimental Medicine* 40, 647-660.

Nalbandian, J., Freedman, M.L., Tanzer, J.M. and Lovelace, S. M. (1974). Ultrastructure of mutants of *Streptococcus mutans* with reference to agglutination, adhesion and extracellular polysaccharide. *Infection and Immunity* 10, 1170-1179.

Ninham, B.W. and Richmond, P. (1973). Multimolecular adsorption on cell surfaces under the influence of van der Waals forces. *Journal of the Chemical Society. Faraday Transactions II* 69, 658-664.

Olson, G.A., Bleiweiss, A.S. and Small, P.A. (1972). Adherence inhibition of *Streptococcus mutans:* an assay reflecting a possible role of antibody in dental caries prophylaxis. *Infection and Immunity* 5, 419-427.

Olson, G.A., Guggenheim, B. and Small, P.A. (1974). Antibody mediated inhibition of dextran/sucrose induced agglutination of *Streptococcus mutans*. *Infection and Immunity* 9, 273-278.

Orstavick, D. (1974). *In vitro* adherence of streptococci to the tooth surface. *Infection and Immunity* 9, 794-800.

Parsegian, V.A. and Gingell, D. (1972). *Recent Advances in Adhesion*. New York and London : Gordon and Breach.

Prout, R.E. (1976). Human saliva lipid, composition and its possible relation to dental calculus formation. *Journal of Dental Research* 55, (NSID) D 134.

Rolla, G. (1977). Inhibition of adsorption: general considerations. In *Microbial Aspects of Dental Caries*, vol. 2, pp.

309-324. Edited by H.M. Stiles, W.J. Loesche and T.C. O' Brien. London : Information Retrieval Ltd.

Rutter, P.R. and Abbott, A. (1978). A study of the interaction between oral streptococci and hard tissues. *Journal of General Microbiology* 105, 219-226.

Rutter, P.R. and Leech, R. (1977). Unpublished work.

Saxton, C.A. (1975). The Formation of Human Dental Plaque. A study of Electron Microscopy. Thesis for Master of Philosophy in the Faculty of Medicine. University of London.

Saxton, C.A. (1976). The effects of dentifrices on the appearance of the tooth surface observed with the scanning electron microscope. *Journal of Periodontal Research* 11, 74-85.

Schachtele, C.F., Harlander, S.K., Fuller, D.W., Zollinger, P. K. and Woon-lam, S.L. (1977). Bacterial interference with sucrose dependent adhesion of oral streptococci. In *Microbial Aspects of Dental Caries*, vol. 2, pp. 401-412. Edited by H.M. Stiles, W.J. Loesche and T.C. O'Brien. London : Information Retrieval Ltd.

Schachtele, C.F., Staat, R.H. and Harlander, S.K. (1975). Dextranases from oral bacteria: Inhibition of water insoluble glucan production and adherence to smooth surfaces by *Streptococcus mutans*. *Infection and Immunity* 12, 309-317.

Socransky, S.C., Manganiello, A.D., Propas, D., Oram, V. and van Houte, J. (1977). Bacteriological studies of developing supragingival dental plaque. *Journal of Periodontal Research* 12, 90-106.

Sonju, T. and Skjorland, K. (1977). Pellicle composition and initial bacterial colonisation on composite and amalgum *in vitro*. In *Microbial Aspects of Dental Caries*, vol. 1, pp. 133-141. Edited by H.M. Stiles, W.J. Loesche and T.C. O'Brien. London : Information Retrieval Ltd.

Tanzer, J.M., Wood, W.I. and Krichevsky, M.I. (1969). Linear growth kinetics of plaque-forming streptococci in the presence of sucrose. *Journal of General Microbiology* 58, 125-133.

Thom, D., Morris, E.R., Rees, D.A. and Welsh, E.J. (1977). Conformation and Intermolecular Interactions of carbohydrate chains. Given at the *6th International ICN - UCLA Symposium on Molecular and Cellular Biology, Colorado, USA*.

Uno, H. and Tanaka, S. (1972). Adhesion of suspension particles on the wall surface of a container. Mechanism of particle adhesion. *Kolloid-Zeitshrift und Zeitshrift für Polymere* 250, 238.

van Oss, C.J. and Gillman, C.F. (1972). Phagocytosis as a surface phenomenon. 1. Contact angles and phagocytosis of non-opsonized bacteria. *Journal of the Reticuloendothelial Society* 12, 283-292.

van Houte, J. and Green, P.B. (1974). Relationship between the concentration of bacteria in saliva and the colonisation of teeth in humans. *Infection and Immunity* 9, 624-630.

Williams, R.C. and Gibbons, R.J. (1975). Inhibition of Streptococcal attachment to receptors on human buccal epithelial cells by antigenically similar salivary glycoproteins. *Infection and Immunity* 11, 711-718.

BACTERIAL ADHESION IN HOST/PATHOGEN INTERACTIONS IN ANIMALS

J.P. ARBUTHNOTT* and C.J. SMYTH**

**Department of Microbiology, Trinity College, Dublin, 2, Ireland, and **Department of Bacteriology and Epizootology, Swedish University of Agricultural Sciences, College of Veterinary Medicine, Biomedical Centre, Uppsala, Sweden.*

INTRODUCTION

Pathogenicity is defined as the ability of microorganisms to cause disease. The aim of this article is to indicate how adhesion of pathogens to host tissue contributes to pathogenicity and to review what is known of the mechanisms of adherence. There is now considerable evidence to show that several pathogenic microorganisms have the ability to adhere to host cells. In some cases it is clear that this is an important factor in pathogenicity but in general the biochemistry of the process is poorly understood. It should be stated that an understanding of the biochemical basis of adherence requires knowledge of the surface components(s) on the pathogen responsible for adherence and knowledge of the corresponding receptor(s) on the tissue cells of the host.

With few exceptions pathogenicity is rarely the property of a single determinant of the pathogen being more usually multifactorial [Smith, 1977]. This can be readily appreciated on considering the following requirements of a pathogen. It must be able to colonise and/or penetrate the body surfaces (Figure 1), namely, the skin or the mucosal epithelia of the respiratory tract, oropharynx, urogenital tract or conjuctiva and it must be capable of multiplying on or in the tissues of the host. Entry into the host may be accompanied by spread through the tissues in which case the pathogen must be able to resist or at least to avoid stimulating the host defences. Finally the symptoms of disease are due to the ability of the pathogen to damage host tissues. Thus the outcome of the interaction between host and the invading microbe depends both on a combination of virulence factors produced by the patho-

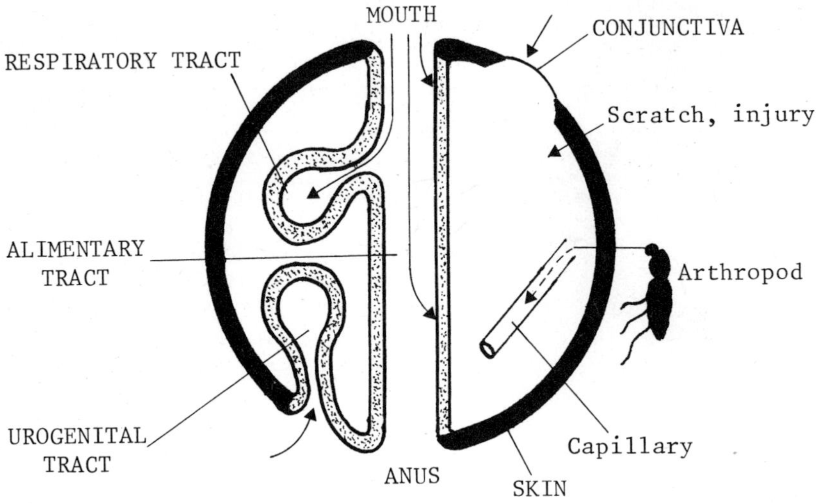

Fig. 1. Body surfaces as sites of infection. Modified from Mims (1977).

gen, and the susceptibility of the host.

Within the complex series of events that comprise the pathogenic process, it is possible to identify the events in which adherence might be important. The first of these is the promotion of attachment to host epithelial surfaces thus giving resistance to the mechanical flushing actions, for example, of moving lumen components, thereby allowing the pathogen to survive and multiply on these surfaces in competition with commensal microorganisms. The other is the promotion of attachment to target tissues within the host at a point distant from the point of entry.

The relative importance of these factors depends on whether the organism exerts its pathogenic effect locally at epithelial surfaces or is primarily an invasive pathogen which penetrates the epithelial surface and spreads through the tissues. Most is known about adherence in relation to the attachment of pathogens to epithelial surfaces (Table 1). Because of limitations of space this review has been restricted to a coverage of bacteria, mycoplasmas and chlamydiae. The role of adherence in the formation of bacterial accumulations on teeth is dealt with in detail in a separate chapter (see Rutter, this volume).

It must be emphasised at the outset that adherence is not an exclusive character of pathogenic microorganisms (see the

Table 1.
Examples of specific adherence of pathogens to mucosal epithelia

Organism	Disease	Site of attachment
Escherichia coli (enteropathogenic strains)	Diarrhoea	Epithelium of small intestine
Vibrio cholerae	Cholera	Epithelium of small intestine
Streptococcus pyogenes	Pharyngitis	Pharyngeal epithelium
Neisseria gonorrhoeae	Gonorrhoea	Urethral and cervical epithelium
Mycoplasma pneumoniae	Atypical pneumonia	Ciliated respiratory epithelium
Mycoplasma (T strains)	Non-gonococcal urethritis	Urethral and cervical epithelium
Chlamydia trachomatis	Conjunctivitis Non-gonococcal urethritis	Conjunctival epithelium Urethral epithelium
Adenovirus	Conjunctivitis Pharyngitis Respiratory illness	Various mucosal epithelia
Influenza virus	Influenza	Bronchia and tracheal epithelium

contributions to this volume by Burns and Fletcher), and thus is not necessarily an indicator of pathogenicity. Indeed, adhesion of indigenous microorganisms to mucosal epithelia in animals (Table 2) plays an important role in the maintenance of the 'normal' flora associated with such ecosystems [Savage, 1975, 1976; Gibbons, 1975; Rutter, this volume]. Thus in order to establish themselves on mucosal surfaces many pathogenic microorganisms must compete successfully with the normal flora [Hentges, 1975; Freter, 1974]. This competition may be

Table 2.
Specific associations between indigenous microorganisms and gastrointestinal mucosal epithelia of rats and mice

Site of attachment	Organisms	Bacterial factors involved in attachment
Stomach (non-secreting region)	*Lactobacillus*	Acid Mucopolysaccharides[1]
Stomach (secreting region)	*Torulopsis*	Mucopolysaccharide[2]
Small intestine	Segmented filamentous bacteria	Specialised attachment[3]
Colon	Fusiform-shaped bacteria	Filaments of unknown composition[4] ?Pili
Colon	Rod-shaped bacteria	Specialised attachment site at one end of cell - involves filaments[5]

[1]Brooker and Fuller [1975]. [2]Savage [1972]. [3]Davis and Savage [1976]. [4]Savage and Blumershine [1974]. [5]Wagner and Barrnett [1974].

due to phenomena such as bacterial interference [Shinefield *et al*., 1972].

ADHESION OF ENTEROPATHOGENIC *ESCHERICHIA COLI* TO INTESTINAL MUCOSA

Certain enteropathogenic strains of *Escherichia coli* (EPEC) are associated with diarrhoeal diseases in pigs, calves, lambs and humans. Despite the differences between strains isolated from different animal species there are two general requirements for the production of disease:

(a) *Colonisation of the small intestine*. Successful colonisation of the small intestine depends on the ability of EPEC to adhere to the mucosal epithelium. This allows the organisms to multiply to reach large numbers and effectively combats the flushing action of peristaltic movements of the intestine. Such colonisation is not usually associated with damage to the normal villous surfaces although a few strains are invasive and can penetrate the mucosal surface and cause local tissue damage.

(b) *Production of enterotoxin(s).* The production of enterotoxins(s) that act on the membranes of mucosal epithelia of the small intestine causes a transmembrane efflux of fluid into the lumen of the gut causing the clinical symptoms of the disease.

In the last ten years both of these aspects of *Escherichia coli* diarrhoea have been intensively investigated and in some strains pilus-like structures on the bacterial surface responsible for adhesion *(adhesins)* have been identified and partially characterised (Table 3). It should be emphasised, however, that characterisation of the heat-labile and heat-stable enterotoxins produced by EPEC has been central to an understanding of the overall pathogenic mechanism. The heat-labile enterotoxins of *Escherichia coli* resemble cholera enterotoxin and activate adenylate cyclase in the membranes of epithelial cells causing the net movement of water and certain ions into the gut lumen [Finkelstein, 1976].

Properties of adhesins in EPEC and the role of adhesion in pathogenicity

Information about adherence and the properties of individual adhesins in EPEC has been derived from several lines of investigation. These include: serotyping of isolates from diarrhoeal disease [Sojka, 1965; Ørskov *et al.*, 1975; Ørskov *et al.*, 1976; Ørskov and Ørskov, 1977]; enumeration of organisms in the small intestine of naturally and experimentally infected animals [Smith and Linggood, 1971; Jones and Rutter, 1972, Bertschinger, Moon and Whipp, 1972; Hohmann and Wilson, 1975; Nagy, Moon and Isaacson, 1976; Evans *et al.*, 1975; Rowland and McCollum, 1977]; fluorescence microscopy and transmission electron microscopy of the mucosal surface of infected animals [Drees and Waxler, 1970; Arbuckle, 1970; Bertschinger *et al.*, 1972; Jones and Rutter, 1972; Hohmann and Wilson, 1975; Nagy *et al.*, 1976; Arbuckle, 1976; Moon, Nagy and Isaccson, 1977]; *in vitro* studies of the adherence of EPEC to intestinal brush border preparations and erythrocytes of certain species [Jones and Rutter, 1974; Gibbons, Jones and Sellwood, 1975; Sellwood *et al.*, 1975; Burrows, Sellwood and Gibbons, 1976]; identification of pilus-like structures by direct electron microscopic examination of the surface of EPEC [Stirm *et al.*, 1967b; Burrows *et al.*, 1976; Evans, Evans and DuPont, 1977]. It is important to emphasise, however, that only the K88 and K99 adhesins of animal strains have been isolated and partially characterised [Stirm *et al.*, 1967a; Isaacson, 1977].

The fact that the production of K88 and K99 antigens, col-

Table 3. *Adhesins associated with enteropathogenic Escherichia coli strains*

Characteristic	K88 Antigen[1]	K99 Antigen[2]	Colonisation Factor[3]	Adherence Factor[4]
Origin of strains causing diarrhoea	Piglets	Calves, lambs	Man	Man
Genetic control	Plasmid	Plasmid	Plasmid	Plasmid
Electron microscopic morphology	Pilus-like layer of flexible filaments	Pilus-like, individual fibrous rods, mean length 120nm, mean diam. 8.4nm	Pilus-like	?(Lipo)polysaccharide
Chemical nature	Protein, 4% lipid	Protein, 6.6% lipid, 0.6% carbohydrate	?	?
Physical properties:				
Sedimentation coefficient (S_{20_w})	36.7S	13-15S	?	?
Molecular weight of SDS subunits	23,000	22,500 and 29,500	?	?
Isoelectric point	?	10.0	?	?

Haemagglutinating properties:				
Erythrocyte species	Guinea-pig	Sheep	Human	see note [5]
Inhibition by monosaccharides	Mannose-insensitive	Mannose-insensitive	Mannose-insensitive	
Absence on bacteria grown at 18°	+	+	?	?
Heat-lability	+	+	+	?
Adhesive properties for intestinal brush borders	+ (piglet)	+ (calf)	+ (rabbit)	+ (human foetal)

[1]Sojka [1965]; Ørskov and Ørskov [1966]; Stirm *et al.* [1967a,b]; Jones [1975]; Jones and Rutter [1974]; Gibbons *et al.* [1975]; Sellwood *et al.* [1975].
[2]Smith and Linggood [1972]; Ørskov *et al.* [1975]; Isaacson [1977]; Burrows *et al.* [1976]
[3]Evans *et al.* [1975; 1977]; Ørskov and Ørskov [1977]; Evans, Evans and Tjoa [1977].
[4]McNeish *et al.* [1975]; Williams *et al.* [1977].

[5]Haemagglutinating activity mannose sensitive but adhesion to human foetal brush borders mannose insensitive.

onisation factor, and adherence factor is controlled by transmissible plasmids [Ørskov and Ørskov, 1966; Smith and Linggood, 1972; Evans *et al.*, 1975; McNeish *et al.*, 1975] has greatly facilitated the study of these adhesins. Also, it has been shown that production of K88 and K99 antigens and colonisation factor depends on the cultural conditions [Jones and Rutter, 1974; Ørskov *et al.*, 1975; Guinée, Jansen and Agterberg, 1976]. Thus, by genetically transferring the appropriate plasmid from strains of EPEC into a suitable recipient such as *Escherichia coli* K12 and/or by manipulating growth conditions it has been possible to correlate adherence with the appearance of pilus-like surface structures and to correlate adhesin production with pathogenicity, with adherence of EPEC to intestinal brush borders and with haemagglutinating activity.

As Table 3 shows K88 and K99 antigens have been more fully investigated than the adhesins of human strains. In part this is due to the additional constraints involved in studying the pathogenicity of human strains. Other factors, however, have influenced work with these strains. For instance the involvement of EPEC in diarrhoeal disease in *adult* humans has been established only recently [Sack *et al.*, 1971; Evans and Evans, 1973; Shore *et al.*, 1974; Sack *et al.*, 1975; Gorbach *et al.*, 1975]. Also there has been a long standing controversy surrounding the aetiological role of so-called 'classical' serotypes in human diarrhoeal disease [Ørskov *et al.*, 1976; Gangarosa and Merson, 1977].

With these reservations in mind it is worth emphasising the similarities between known adhesins in EPEC. Each is a heat-labile, pilus-like surface antigen coded for by a transmissible plasmid, (see Figure 2 for example). In addition to their adhesive properties for intestinal brush borders they cause mannose-insensitive haemagglutination of mammalian erythrocytes; this latter characteristic distinguishes them from common fimbriae [Duguid, 1968]. The receptors for these adhesins in intestinal mucosal cells have not been identified although indirect evidence based on studies of inhibition of haemagglutination suggest that β-D-galactosyl residues in heterosaccharide side chains of membrane glycoproteins might be involved [Gibbons, Jones and Sellwood, 1975].

Experimental evidence supporting the role of K88 antigen in the pathogenicity of EPEC is convincing [Smith and Linggood, 1971; Jones and Rutter, 1972; Rutter *et al.*, 1976]. The adhesin is produced *in vivo* in the small intestine and its production allows K88-positive strains to adhere to and colonise that site but not the large intestine. Also, piglets phenotypically lacking the receptor for K88-antigen [Sellwood *et al.*, 1975] or piglets suckled by dams vaccinated with K88 vaccine [Rutter *et al.*, 1976] are not susceptible to challenge

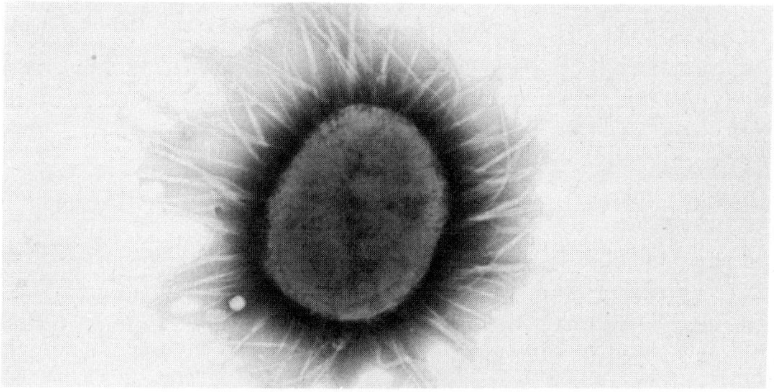

Fig. 2. Negatively stained preparation of an enteropathogenic human strain of Escherichia coli that produces colonisation factor. Note the numerous pilus-like surface structures. Kindly supplied by Dr. D.G. Evans, Texas Medical School.

with K-88-positive EPEC. Such direct evidence is lacking for K99 antigen and colonisation factor although available indirect evidence strongly suggests that these adhesins are determinants in pathogenicity.

However, it should be emphasised that not all strains of EPEC isolated from pigs [Bertschinger *et al.*, 1972; Hohmann and Wilson, 1975; Nagy *et al.*, 1976] produce K88 antigen and it has recently been shown that production of colonisation factor by human EPEC strains is restricted to serotypes containing the O78 antigen [Ørskov and Ørskov, 1977]. Clearly other adhesins remain to be identified and it has been suggested that polysaccharide K antigens and pili may be involved in colonisation of the small intestine by K88-negative porcine strains of EPEC [Isaacson, Nagy and Moon, 1977; Nagy, Moon and Isaacson, 1977; Moon, Nagy and Isaacson, 1977].

The view that adherence to the small intestine is a prerequisite for successful colonisation by enteropathogenic *Escherichia coli* has been questioned recently by Arbuckle [1976]. On the basis of ultrastructural and bacteriological investigations, he concluded that proliferation and adherence were independent of each other and that proliferation might be promoted by a reduction of intestinal motility preceding the onset of diarrhoea.

ADHESION OF VIBRIO CHOLERAE

As in the case of *Escherichia coli* diarrhoea, the onset of cholera, that is of fluid accumulation in the intestinal lumen caused by activation of adenylate cyclase in the intestinal mucosa by cholera enterotoxin, is concomitant with the establishment of large number of organisms in the intervillous spaces and crypts of the intestine after penetration of the mucus layer between the intestinal contents and the tips of the villi by *Vibrio cholerae* [Finkelstein, 1973; Holmgren and Svennerholm, 1977].

Studies on the kinetics of colonisation of intestinal villi using scanning and transmission electron microscopy have revealed new information on the time course of the adhesion, multiplication and fluid accumulation phases of the infection [Guentzel and Berry, 1975; Schrank and Verwey, 1976; Nelson, Clements and Finkelstein, 1976]. A comprehensive series of papers has recently dealt with the adhesive phase of cholera [Jones, Abrams and Freter, 1976; Freter and Jones, 1976; Jones and Freter, 1976]. These workers used three *in vitro* model systems, namely attachment to rabbit intestinal brush border membranes, haemagglutination of human erythrocytes and attachment to slices of rabbit ileal intestinal mucosa (Table 4). Such studies illustrate the value of *in vitro* systems but also exemplify the limitations of extrapolating to the *in vivo* situation, from observations made in a single *in vitro* system.

The motility of vibrios has often been proposed as a contributing factor to virulence [Williams *et al.*, 1973; Guentzel and Berry, 1975; Nelson *et al.*, 1976]. Indeed the studies of Guentzel and Berry [1975] have demonstrated that non-motile mutants given by oral challenge to suckling infant mice possessed much reduced virulence and reduced adhesive capacity towards mouse intestinal segments *in vitro*. The Ann Arbor group could, however, discriminate between motility and adhesive properties *in vitro* in the brush border and haemagglutination models, but not in the intestinal mucosal system (Table 4).

Using an approach earlier developed by Gibbons *et al.* [1975] (see the previous main section) L-fucose was shown to inhibit adhesion of *Vibrio cholerae* organisms to brush borders and erythrocytes, but not adherence to intestinal slices (Table 4) [Freter and Jones, 1976]. Moreover, adherence to intestinal mucosa differed in other respects from the adhesive properties observed in the other models. There was a requirement for calcium ions, a spontaneous elution of attached vibrios, inhibition by D-mannose, and inhibition by antibodies to the somatic O antigen of *Vibrio cholerae* (Table 4). A soluble inhibitor from pepsin digests of heated intestinal mucosa scrapings functioned in all three models.

Table 4. *Comparative adhesive properties of Vibrio cholerae in three in vitro attachment model systems. Adapted from Freter and Jones (1976)*

Characteristic	Adhesion to rabbit intestinal brush borders	Haemagglutination of human erythrocytes	Attachment to rabbit ileal intestinal mucosa
Bacterial phenotype/genotype:			
agar-grown (motile)	-	-	+
broth culture (motile)	+	+	+
non-motile mutants (Flagella⁻)	-	-	-
motile revertants	+	+	+
Spontaneous elution of attached broth culture vibrios	+	+	-
Attachment in the absence of added calcium ions	-	-	+
Inhibition of attachment:			
by L-fucose	+	+	-
by D-mannose	±	-	-
by anti-vibrio O-antibody	-	-	+
by intestinal mucosal extract (pepsin-digest)	+	+	+

There appear to be at least two specific mucosal receptors for *Vibrio cholerae*, one of which is fucose-sensitive on the brush border epithelium, whereas the other is fucose-resistant and of unknown microanatomical location on the intact mucosa. These factors would not have been revealed without the use of *all three* systems. Motility appears to play a role in the transport of vibrios to their fucose-resistant receptors and to facilitate penetration of the intestinal mucus layer [Jones *et al.*, 1976] but adhesion does not seem to depend only on motility. Phenotypic expression of the fucose-sensitive adhesin appears to depend on environmental factors (such as the medium) at least *in vitro* (Table 4).

It should be remembered that vibrio flagella are sheathed structures [Follett and Gordon, 1963]. Naked (unsheathed) flagella have not proved to be highly protective as a vaccine against oral challenge with virulent vibrios in the suckling mouse model, whereas crude flagella provided complete protection [Eubanks, Guentzel and Berry, 1977]. The latter observations suggested that sheath material may be the protective antigen, the flagellar sheath may thus be a carrier of one adhesin. However, the studies of Freter and Jones [1976] appear to indicate that the O somatic antigen may also be important in adhesion. This conclusion would be in agreement with electronmicroscopic observations that the vibrios attach to epithelium terminally with the flagella directed outwards into the mucus and subsequently by contact between the cell body and the microvilli [Nelson *et al.*, 1976].

The possibility that adhesion of *Vibrio cholerae* may be multifactorial with involvement of pilus-like structures [Wilson *et al.*, 1976] and/or an extracellular slime layer [Lankford and Legsomburana, 1965] remains to be established.

ADHESION OF *STREPTOCOCCUS PYOGENES*

The common pathogen *Streptococcus pyogenes* (group A streptococcus) is responsible for many infections including pharyngitis (tonsilitis or sore throat). The organism is therefore able to colonise and proliferate on the epithelial cells of the oral cavity. The mechanism of adherence of group A streptococci to epithelial cells has recently been the subject of several investigations. Some progress has been made in identifying the streptococcal surface components responsible for adhesion but nothing is known of the chemical nature of receptors on tissue cells.

The surface of group A streptococci as revealed by electron microscopy consists of a layer of fine projections referred to as 'fimbriae' or surface 'fuzz'. This surface layer contains the type specific M protein antigen. M proteins are extractable [Fox, 1974], the traditional acid-extractable material yielding a complex mixture of partially degraded polypeptides ranging in molecular weight from 20,000 to 40,000; less drastic methods (extraction with alkali, detergents or lytic enzymes) yield a less complex mixture of polypeptides of higher molecular weight.

Other surface structures of *Streptococcus pyogenes* include the hyaluronic acid capsule, the T protein, serum opacification factor, M-associated protein and lipoteichoic acid. The surface of this organism can thus be envisaged as a mosaic of antigenic components arranged within a three-dimensional reticulum of 'fimbriae' or 'fuzz'. The work of Ellen and Gibbons [1972] indicated that the ability of Group A streptococci to adhere to human oral epithelial cells correlated with possession of M-protein.

In an attempt to obtain information about the specificity of adherence of group A streptococci to epithelial cells, the adhesion of an M-positive strain of *Streptococcus pyogenes* and an enteropathogenic strain of *Escherichia coli* to several types of mucosal surfaces in germ free rats was studied [Ellen and Gibbons, 1974]. Adherence of the *Streptococcus pyogenes* strain to the tongue surface was six times greater than the *Escherichia coli* strain. As part of the same study it was observed that adherence of *Streptococcus pyogenes* was greatest in early stationary phase cells and that many different treatments inhibited adhesion (trypsin, lipase, surfactants, heating to 50°C and exposure to phospholipids). These results prompted a note of caution in the interpretation of the mechanism of adhesion and Ellen and Gibbons concluded that although M protein was involved it might not be solely responsible for actual binding to epithelial cells.

The role of M protein in adhesion of group A streptococci has been questioned by Ofek's group [Ofek *et al.*, 1975; Beachy and Ofek, 1976] who proposed that small amounts of lipoteichoic acid (polyglycerophosphate ester-linked to fatty acids) are exposed on the surface of group A streptococci and that the binding of 'fimbriae' to epithelial cells involves lipoteichoic acid rather than M protein.

Purified lipoteichoic acid adsorbed to human erythrocytes making them agglutinable to anti-lipoteichoic acid antiserum and chemical evidence suggested that ester-linked fatty acids were involved in the adsorption process [Ofek *et al.*, 1975]. Fimbriate surface structures and lipoteichoic acid were unaffected by treatment with pepsin at pH 5.8 which selectively removed M protein [Beachy and Ofek, 1976]. Streptococci treated in this way retained their ability to adhere to epithelial cells and experiments with purified antigens and specific antisera supported the view that lipoteichoic acid rather than M protein was the adhesin of group A streptococci.

ADHESION OF *NEISSERIA GONORRHOEAE*

In vitro cultivation of *Neisseria gonorrhoeae* leads to dissociation of the organisms into four so-called Kellogg colony types termed T1 to T4 [Kellogg *et al.*, 1963; McEntegart, 1975]. Types T1 and T2 were shown in large doses to produce clinical gonorrhoea in human volunteers, whereas T3 and T4 were avirulent [Kellogg *et al.*, 1963]. These observations and the subsequent demonstration by electron microscopy of pilus-like structures on the surface of *in vitro*-grown gonococci of these putatively virulent Kellogg types [Smith, 1977] stimulated research on the adherence of gonococci [Swanson, Kraus and Gotschlich, 1971; Jephcott, Reyn and Birch-Andersen, 1971]. Indeed, pili have been purified from *in vitro*-grown gonococci in several laboratories and chemically characterised [Buchanan *et al.*, 1973; Buchanan, 1975; Reimann and Lind, 1977; Robertson, Vincent and Ward, 1977].

Observations demonstrating greater adhesion of piliated than non-piliated gonococci to different tissue culture cells are now numerous, for example, see Ward and Watt [1975] and Swanson [1977]. Although Heckels *et al.* [1976] considered that the importance of pili is to overcome the electrostatic barrier to attachment they were unable to demonstrate that pili had a lower charge density than the surrounding bacterial envelope. If such differences in charge density exist they could hypothetically facilitate anchoring of the gonococci to the negatively charged host membrane by overcoming electrostatic repulsion.

Anti-pilus antibodies have been detected in sera from patients with gonorrhoea, indicating their presence immunogenically *in vivo* [Buchanan *et al.*, 1973]. Furthermore, the preliminary observations of Brinton [cited by Robertson *et al.*, 1977] indicated that immunisation with purified pili increased the resistance of human volunteers to challenge with *in vitro*-grown organisms. Despite a considerable body of evidence, the role of pili in gonococcal attachment *in vivo* during natural infection remains an area of controversy.

A non-pilus surface factor termed "leucocyte association factor" which markedly affected the association of gonococci with leucocytes has been described by Swanson *et al.* [1974]. Also Tramont and Wilson [1977] have demonstrated considerable variation in the adhesive properties of T1 colony types (piliated) from different clinical isolates. Thus, *in vitro* evidence suggests that factors other than pili are involved in gonococcal attachment. However, the central controversy surrounds the comparative properties of *in vitro*-grown organisms, *in vivo*-grown organisms (grown in chambers in animals) and gonococci in pathological specimens (now commonly referred to as host-grown gonococci). In general, observations on *in vivo* and host-grown gonococci cast doubt on the primary importance of pili (as defined as morphological entities on *in vitro* grown cells) in the pathogenesis of the natural infection.

These findings can be summarised as follows: Electron-microscopic observations showed that host-grown organisms only occasinally possessed pili; bundles of pili typical of *in vitro*-grown organisms (Figure 3A) have not been identified and thin sections of gonococci adhering to host-derived mucosal cells did not reveal pilus-like structures [Ward and Watt, 1975; Novotny, Short and Walker, 1975; Ward *et al.*, 1975; Watt, Ward and Robertson, 1976; Evans, 1977; Watt and Ward, 1977]. The surface layers of *in vivo* grown and host-grown gonococci differ considerably from *in vitro*-grown organisms [Novotny *et al.*, 1975; Arko, Bullard and Duncan, 1976] (see Figure 3B and 3C). These morphological differences may be related to observed changes in antigenic and biological properties [Ward, Watt and Glynn, 1970; Arko *et al.*, 1976; Penn, Veale and Smith, 1977; Penn *et al.*, 1976, 1977]. Novotny *et al.* [1977] propose that the gonococci capable of initiating infection attach to epithelial cells in so-called 'infectious units', that is clusters of multiplying gonococci contained within vesicular membranes derived from damaged macrophages; this is supported by the recent work of Evans [1977]. Thus, direct interaction between gonococcal adhesins and the mucosal surface [Ward and Watt, 1975; Ward *et al.*, 1975; Tebbutt *et al.*, 1976, 1977; Watt *et al.*, 1976; Evans, 1977; Watt and Ward, 1977] may represent only one type of adhesive interaction in the natural infection.

Gonococcal research has expanded dynamically in the past seven years, though the mechanism of gonococcal pathogenicity is complex and difficult to study. However, it is generally accepted that adhesion is an essential stage in allowing the pathogen to become established and proliferate on mucosal surfaces. It is impossible at the present time to make a definitive statement concerning the role of a pilus-associated adhesin, especially as such a putative adhesin could possibly exist in a different microanatomical form on *in vivo*-grown and host-grown organisms [Swanson, 1972; Novotny *et al.*, 1975; Hodgkiss, Short and Walker, 1976]. As yet, little is known about the tissue receptor(s) involved in host and tissue specificity *in vivo*.

ADHESION OF MYCOPLASMAS AND CHLAMYDIAE

Mycoplasmas

Pathogenic mycoplasmas possess adhesive properties for

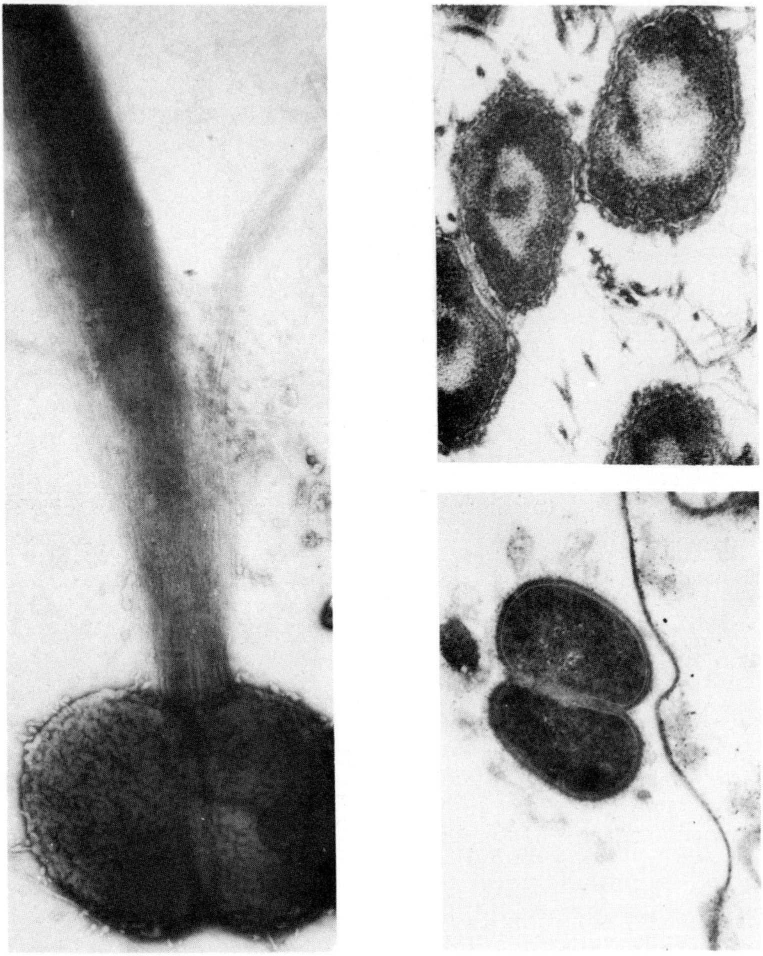

Fig. 3. A (left): Negatively stained preparation of Neisseria gonorrhoeae (strain Pat 1) harvested from liquid GC medium in exponential growth phase showing bundles of pili in the region of the septum (x 30,600). B (top right): Thin section of Neisseria gonorrhoeae (strain Pat 1) from an 18 hour culture on solid medium showing the "crumpled" appearance of the outer membrane and tangentially sectioned pili (x 36,000). C (bottom right): Thin section of Neisseria gonorrhoeae in pus from acute gonorrhoea ("host-grown organisms") showing the smooth structure of the envelope layers (x 36,000). Photograph kindly supplied by Dr. P. Novotny.

cell membranes and inert surfaces [Clyde, 1975]. Study of the attachment of *Mycoplasma pneumoniae* to host cells has been facilitated by the use of hamster and human tracheal organ cultures [Collier and Baseman, 1973]. In its natural infective state *Mycoplasma pneumoniae* is a single filamentous microorganism that attaches to ciliated tracheal epithelial cell surfaces by means of a differentiated tip structure (organelle) [Collier and Clyde, 1971]. The model organ culture system reflects the cell biology of the natural infection [Collier and Clyde, 1974] in which the mycoplasmas localise on the microvilli of the respiratory epithelium and between the cilia. Attachment *in vitro* occurs rapidly despite the mucous layer and the synchronous beating of the cilia though how the organism penetrates the mucous barrier is unknown. Avirulent strains of *Mycoplasma pneumoniae* have limited attachment properties.

Sialyl residues appear to be associated with the receptors on tracheal epithelial cells on the basis of enzyme treatments and inhibition of attachment by sialic acid [Sobeslavsky, Prescott and Chanock, 1968; Collier and Baseman, 1973; Powell *et al.*, 1976; Gorski and Bredt, 1977]. Most of the evidence from these authors suggests that the adhesin(s) of *Mycoplasma pneumoniae* is (are) protein in nature. Interestingly, Gabridge *et al.* [1977] have reported that the mechanism of attachment of *Mycoplasma pneumoniae* membranes to respiratory epithelium is distinct from the adhesin-mediated attachment of intact cells. This observation emphasises the importance of the differentiated tip structure.

Other mycoplasmas have differentiated organelles that relate to their adhesiveness [Clyde, 1975; Allan and Pirie, 1977]. Sialic acid-containing receptors have been shown to play a role in *Mycoplasma gallisepticum*-mediated haemagglutination [Gesner and Thomas, 1965]. *Mycoplasma hominis, Mycoplasma salivarium* and *Mycoplasma dispar* appear to attach to neuraminidase-resistant cell receptors [Manchee and Taylor-Robinson, 1969]. Indeed, the results of Hollingdale and Manchee [1972] support the suggestion that the receptors for *Mycoplasma hominis* are protein in nature.

Chlamydiae

Chlamydiae are obligate intracellular parasites. Attachment to host cell membranes is thus an integral step in their life-cycle. This group of organisms is currently receiving increased attention with the realisation that they cause a genital tract infection referred to as non-gonococcal urethritis [Grayston and Wang, 1975; Hobson and Holmes, 1977]. The *Chlamydia trachomatis* organisms causing trachoma and non-gonococcal urethritis and lymphogranuloma venereum could be differentiated on the basis of their attachment properties to tissue culture monolayers [Kuo, Wang and Grayston, 1972, 1973; Kuo and Grayston, 1976].

Sialic acid residues on the surface of host cells may act as receptors for trachoma organisms at least under conditions of centrifuge-assisted infection, that is enhanced infection of cells by centrifugation of the inoculum onto the cell monolayers [Weiss and Dresler, 1960].

Recent observations on the susceptibility of cell lines to infection by a strain of *Chlamydia psittaci* that causes guinea-pig inclusion conjunctivitis suggest that serum modulation of the cell surface affects adhesion [Allan and Pearce, 1977; Allan, Spragg and Pearce, 1977]. These authors also concluded that centrifuge-assisted infection depended on induction of a host cell-membrane response, for example the pressure-induced deformation of the cell surface, which was absent in spontaneous chlamydial infection of static cell monolayers. It is thus questionable whether sialic acids residues, putatively assigned receptor function under experimental conditions utilising centrifuge-assisted infection, represent naturally exposed receptors or those induced by pressure effects. Also, attachment does not appear to be the only factor that determines the susceptibility or resistance of cell lines to infection [Friis, 1972].

ADHESION OF MISCELLANEOUS BACTERIA

Studies on other putative bacterial adhesions are tabulated in Table 5. No pretence at completeness is intended, but the examples demonstrate some important deficiencies in our knowledge. Pili have often been claimed to play a role in attachment, but definitive evidence to justify such claims is still lacking in most instances.

A criterion which should be fulfilled is whether or not the association of bacteria with a particular tissue is specific or non-specific, that is whether selective adherence to the host tissue can be demonstrated. This is illustrated in the studies of Frost [1975] and Frost, Wanasinghe and Woolcock [1977]. The adhesiveness of *Staphylococcus aureus* and *Streptococcus agalactiae*, proven pathogens associated with bovine mastitis, was much greater than that of bacteria such as *Escherichia coli* and *Streptococcus faecalis* which are isolated infrequently from such infections. These continuing studies promise new insight into adherence mechanisms in the pathogenesis of bovine mastitis.

Table 5. *Adhesion of miscellaneous bacteria to host tissues*

Bacterium	Infection	Proposed bacterial adhesin	Reference
Corynebacterium renale	Pyelonephritis in cattle	Pili	Honda and Yanagawa [1974]
Salmonella enteritidis	Enteritis	Unknown. L-fucose insensitive	Freter and Jones [1976]
Escherichia coli	Urinary tract infections	K antigens (K1,K2,K3, K12,K13 polysaccharides)	Kaijser *et al.* [1977] Glynn *et al.* [1971]
		Pili, mannoside-sensitive (type 1?)	Salit and Gotschlich [1976]
		Pili (tenacious surface component), hexose-, glucosamine- and sialic acid-sensitive	Dinh *et al.* [1976]
Salmonella typhimurium	Enteritis	Unknown, but not pili, flagella or O antigen	Tannock *et al.* [1975]
Moraxella bovis	Keratoconjunctivitis in cattle	Pili	Pedersen *et al.* [1972] Pugh and Hughes [1976]
Proteus mirabilis	Pyelonephritis	Pili	Silverblatt and Ofek [1976]
Streptococcus agalactiae, *Staphylococcus aureus*	Bovine mastitis	Pili	Frost [1975]; Frost *et al.* [1977]; Hodgkiss *et al.* [1976]
Clostridium perfringens	Enteritis in pigs	?	Arbuckle [1972]

CONCLUSION

Sequence of events in adhesion

Receptors for microbial adhesins should be considered as sites for which the microbe has physicochemical affinity and with which a stable non-covalent adhesive bond is established [Meager and Hughes, 1977]. Attachment to the surface of a mucosal membrane involves penetration of the protective mucus, though the mechanisms by which bacteria achieve penetration generally lack experimental definition.

Motility in the case of *Vibrio cholerae* [Jones *et al.*, 1976] and probably in the case of motile enteropathogenic *Escherichia coli* provides for mechanical penetration along the planes of least resistance within the intestinal mucus. Sialidases and glycosidases (mucinases) produced by *Vibrio cholerae, Clostridium perfringens* and *Corynebacterium diphtheriae* may create random tracks in the mucus terminating at the cell surface. Similarly, membrane-bound proteases have been invoked to explain penetration of K88-positive *Escherichia coli* [Gibbons *et al.*, 1975]. Sperm with attached gonococci have been suggested to provide vehicular transport of their 'hitch-hikers' through cervical mucus [James-Holmquest *et al.*, 1974]. Moreover it is possible that the pathogen approaches the cell surface in a particular orientation, as a result of response to a chemotactic gradient, possession of specialised adhesive polar structures and propulsion by polar flagella.

Electrostatic repulsive forces over short distances probably represent the initial barrier to interaction between the bacterium and the host cell membrane itself. Devices such as alteration of or reduction in cell surface charge, for example by K99 antigen with a pI > 10.0 [Isaacson, 1977] may overcome the forces of repulsion. Generally this aspect of adhesion has not been explored. Initial contact with the host cell surface presumably involves forces of attraction such as van der Waals and London forces. This initial interaction is probably random. Such collisions may result in high affinity interactions displaying close complementarity between the adhesion and chemical groupings on the host membrane.

The molecular basis of interactions leading to host-pathogen adhesion is unknown, but it seems likely that analogies may exist with antigen-antibody, lectin-oligosaccharide or enzyme-substrate interactions. Indeed, there is some evidence that the interaction between common type 1 pili and erythrocytes or other cell surfaces resembles that of lectins [Old, 1972; Ofek, Mirelman and Sharon, 1977]. Adhesin-mediated attachment may be followed by stronger adhesive interactions involving larger areas of the microbial cell body.

It is envisaged that future investigations of adhesins

should focus on more detailed characterisation (both physical and chemical) of putative adhesins and the use of techniques such as phase partition, hydrophobic interaction and affinity chromatography to study their binding properties.

In Vitro and In Vivo Test Systems

In assessing *in vitro* model systems it is essential to bear in mind that these may not be comparable to each other or to the *in vivo* situations during infection. Several factors will influence the outcome of investigations in model systems. These are firstly that the production of adhesins *in vitro* has been shown to depend on cultural conditions [Guinee *et al.*, 1976; Jones *et al.*, 1976] and differences have been demonstrated between *in vitro* and *in vivo* grown gonococci [Penn *et al.*, 1976; Arko *et al.*, 1976]. Secondly the value of studies involving enzymatic modification of either the host-cell surface or the bacterial adhesin depends on the purity of the enzyme(s) used, for example, see Den, Malinzak and Rosenberg [1975], and lastly that the isolation of epithelial membranes such as brush border membranes from small intestine may result in the loss of loosely attached surface components or modification of the orientation or expression of surface components.

Haemagglutination has been widely used as an index of the adhesiveness of pathogenic microorganisms. Although the convenience of the system is attractive, its relevance can be questioned. It seems unlikely that the relatively simple cell surface of erythrocytes reflects that of mucosal epithelial cells which are physiologically and microanatomically more complex.

In some cases haemagglutination occurs only in the cold [Sellwood *et al.*, 1975] suggesting that the affinity for the receptor in erythrocytes is less good than for the natural receptor. In a comparison of three different *in vitro* models for the adherence of *Vibrio cholerae*, haemagglutinating activity did not parallel adherence to slices of rabbit ileum [Freter and Jones, 1976] or to isolated brush borders [Jones *et al.*, 1976]. Moreover, similar reservations may apply to the use of preparations of target tissue *in vitro*. For instance, K88-negative enteropathogenic *Escherichia coli* have the ability to adhere to the surface of the small intestine *in vivo*, but not *in vitro* [Hohmann and Wilson, 1975]. Thus, where possible model systems should be carefully evaluated in relation to *in vivo* observations.

Adhesins as Protective Antigens

Where adhesins are known to be determinants of pathogenicity, it may be possible to improve methods of protection by preventing the bacteria from colonising host tissues. Such agents would have considerable potential in prophylaxis against certain infectious diseases of man and animals. Although this objective is the long-term aim of much work on bacterial adhesion, in general studies in this field have not reached an advanced stage. To date the most thoroughly tested system for the protective capacity of anti-adhesive activity is that involving K88 antigen [Rutter *et al.*, 1976]. Neutralisation of the adhesive properties of K88 antigen by K88 antibodies in colostrum and milk, contributed significantly to the protection of piglets suckled by vaccinated dams.

Evaluation of surface components of *Vibrio cholerae* [Eubanks *et al.*, 1977] and *Neisseria gonorrhoeae* [Buchanan *et al.*, 1977] as potential protective immunogens is underway. However, the demonstrated antigenic heterogeneity of gonococcal surface antigens, such as pili [Buchanan, 1975; Novotny and Turner, 1975] and outer membrane protein complexes [Johnston, Holmes and Gotschlich, 1976], presents additional problems in assessing their potential promise as vaccine components as neither the total numbers of serotypes nor their prevalences in different strains has yet been assessed.

Coda

The level of understanding of adhesive processes with many bacterial host-cell systems leaves much room for initiative. High resolution crossed immunoelectrophoretic analysis as used for definition of mycoplasmal and chlamydial membranes [Johansson and Hjertén, 1974; Alexander and Kenny, 1977; Caldwell, Kuo and Salton, 1977; Smyth *et al.*, 1978] may prove to be a useful tool for identification, purification and characterisation of bacterial adhesins and their host cell receptors [Bjerrum and Bog-Hansen, 1976; Bjerrum, 1977]. The broadening interest in the role of adhesion in pathogenesis augurs well for future breakthroughs and developments.

REFERENCES

Alexander, A.G. and Kenny, G.C. (1977). Characterization of membrane and cytoplasmic antigens of *Mycoplasma arginini* by two-dimensional (crossed) immunoelectrophoresis. *Infection and Immunity* 15, 313-321.

Allan, E.M. and Pirie, H.M. (1977). Electronmicroscopical observations on mycoplasmas in pneumonic calves. *Journal of Medical Microbiology* 10, 469-472.

Allan, I. and Pearce, J.H. (1977). Serum modulation of cell susceptibility to chlamydial infection. *FEMS Microbiology Letters* 1, 211-214.

Allan, I., Spragg, S.O. and Pearce, J.H. (1977). Pressure and directional force components in centrifuge-assisted chlamydial infection of cell cultures. *FEMS Microbiology Letters* 2, 79-82.

Arbuckle, J.B.R. (1970). The location of *Escherichia coli* in the pig intestine. *Journal of Medical Microbiology* 3, 330-340.

Arbuckle, J.B.R. (1972). The attachment of *Clostridium welchii (Cl. perfringens)* type C to intestinal villi of pigs. *Journal of Pathology* 106, 65-72.

Arbuckle, J.B.R. (1976). Observations on the association of pathogenic *Escherichia coli* with the small intestinal villi of pigs. *Research in Veterinary Science* 20, 233-236.

Arko, R.J., Bullard, J.C. and Duncan, W.P. (1976). Effects of laboratory maintenance on the nature of surface reactive antigens of *Neisseria gonorrhoeae*. *British Journal of Venereal Diseases* 52, 316-325.

Beachy, E. and Ofek, I. (1976). Epithelial cell binding of group A streptococci by lipoteichoic acid on fimbriae denuded of M protein. *Journal of Experimental Medicine* 143, 759-771.

Bertschinger, H.U., Moon, H.W. and Whipp, S.C. (1972). Comparison of enteropathogenic and nonenteropathogenic porcine strains in pigs. *Infection and Immunity* 5, 595-605.

Bjerrum, O.J. (1977). Immunochemical investigation of membrane proteins. A methodological survey with emphasis placed on immunoprecipitation in gels. *Biochemica et Biophysica Acta* 472, 135-196.

Bjerrum, O.J. and Bog-Hansen, T.C. (1976). Immunochemical gel precipitation techniques in membrane studies. In *Biochemical Analysis of Membranes*, pp. 378-426. Edited by A.H. Maddy. London : Chapman and Hall.

Brooker, B.E. and Fuller, R. (1975). Adhesion of lactobacilli to the chicken crop epithelium. *Journal of Ultrastructural Research* 52, 21-33.

Buchanan, T.M. (1975). Antigenic heterogeneity of gonococcal pili. *Journal of Experimental Medicine* 141, 1470-1475.

Buchanan, T.M., Pearce, W.A., Schoolnik, G.K. and Arko, R.J. (1977). Protection against infection with *Neisseria gonorrhoeae* by immunization with outer membrane protein complex and purified pili. *Journal of Infectious Diseases* 136, Supplement, S132-S137.

Buchanan, T.M., Swanson, J., Holmes, K.K., Kraus, S.J. and Gotschlich, E.C. (1973). Quantitative determination of antibody to gonococcal pili. Changes in antibody levels with gonococcal infection. *Journal of Clinical Investigations* 52, 2896-2909.

Burrows, M.R., Sellwood, R. and Gibbons, R.A. (1976). Haemagglutinating and adhesive properties associated with the K99 antigen of bovine strains of *Escherichic coli*. *Jour-*

nal of General Microbiology 96, 269-275.
Caldwell, H.D., Kuo, C.-C. and Kenny, G.E. (1975a). Antigenic analysis of chlamydiae by two-dimensional immunoelectrophoresis. I. Antigenic heterogeneity between *C. trachomatis* and *C. psittaci*. *Journal of Immunology* 115, 963-968.
Caldwell, H.D., Kuo, C.-C. and Kenny, G.E. (1975b). Antigenic analysis of chlamydiae by two-dimensional immunoelectrophoresis. II. A trachoma-LGV-specific antigen. *Journal of Immunology* 115, 969-975.
Clyde, W.A., Jr. (1975). Pathogenic mechanisms in mycoplasma diseases. In *Microbiology - 1975*, pp. 143-146. Edited by D. Schlessinger. Washington : American Society for Microbiology.
Collier, A.M. and Baseman, J.B. (1973). Organ culture techniques with mycoplasmas. *Annals of the New York Academy of Sciences* 225, 277-289.
Collier, A.M. and Clyde, W.A., Jr. (1971). Relationships between *Mycoplasma pneumoniae* and human respiratory epithelium. *Infection and Immunity* 3, 694-701.
Collier, A.M. and Clyde, W.A., Jr. (1974). Appearance of *Mycoplasma pneumoniae* in lungs of experimentally infected hamsters and sputum from patients with natural disease. *American Review of Respiratory Diseases* 110, 765-773.
Davis, C.P. and Savage, D.C. (1976). Effect of penicillin on the succession, attachment and morphology of segmented, filamentous microbes in the murine small bowel. *Infection and Immunity* 13, 180-188.
Den, H., Malinzak, D.A. and Rosenberg, A. (1975). Cytotoxic contaminants in commercial *Clostridium perfringens* neuraminidase preparations purified by affinity chromatography. *Journal of Chromatography* 111, 217-222.
Dinh, T.H., Stalons, D.R. and Swenson, R.M. (1976). Adherence of *Escherichia coli* to urethral mucosal cells in the pathogenesis of urinary tract infections. *Abstracts of Interscience Conference on Antimicrobial Agents and Chemotherapy* 16, No.11.
Drees, D.T. and Waxler, G.L. (1970). Enteric colibacillosis in gnotobiotic swine: a fluorescence microscopic study. *American Journal of Veterinary Research* 31, 1147-1157.
Duguid, J.P. (1968). The function of bacterial fimbriae. *Archivum Immunologiae et Therapiae Experimentalis* 16, 173-188.
Ellen, R.P. and Gibbons, R.J. (1972). M-protein associated adherence of *Streptococcus pyogenes* to epithelial surfaces: prerequisite for virulence. *Infection and Immunity* 5, 826-830.
Ellen, R.R. and Gibbons, R.J. (1974). Parameters affecting

the adherence and tissue tropisms of *Streptococcus pyogenes*. *Infection and Immunity* 9, 85-91.

Eubanks, E.R., Guentzel, M.N. and Berry, L.J. (1977). Evaluation of surface components of *Vibrio cholerae* as protective antigens. *Infection and Immunity* 15, 533-538.

Evans, B.A. (1977). Ultrastructural study of cervical gonorrhoea. *Journal of Infectious Diseases* 136, 248-255.

Evans, D.G., Evans, D.J., Jr. and Dupont, H.L. (1977). Virulence factors of enterotoxigenic *Escherichia coli*. *Journal of Infectious Diseases* 136, Supplement, S118-S123.

Evans, D.G., Evans, D.J., Jr. and Tjoa, W. (1977). Haemagglutination of human group A erythrocytes by enterotoxigenic *Escherichia coli* isolated from adults with diarrhoea : Correlation with colonization factor. *Infection and Immunity* 18, 330-337.

Evans, D.G., Silver, R.P., Evans, D.J., Jr., Chase, D.G. and Gorbach, S.L. (1975). Plasmid-controlled colonization factor associated with virulence in *Escherichia coli* enterotoxigenic for humans. *Infection and Immunity* 12, 656-667.

Evans, D.J., Jr. and Evans, D.G. (1973). Three characteristics associated with enterotoxigenic *Escherichia coli* isolated from man. *Infection and Immunity* 8, 322-328.

Finkelstein, R.A. (1973). Cholera. *Critical Reviews in Microbiology* 2, 553-623.

Finkelstein, R.A. (1976). Progress in the study of cholera and related enterotoxins. In *Mechanisms of Bacterial Toxinology* pp. 53-84. Edited by A.W. Bernheimer. New York and London : John Wiley and Sons, Inc.

Follett, E.A.C. and Gordon, J. (1963). An electron microscope study of vibrio flagella. *Journal of General Microbiology* 32, 235-239.

Fox, E.N. (1974). M proteins of group A streptococci. *Bacteriological Reviews* 38, 57-86.

Freter, R. (1974). Interactions between mechanisms controlling the intestinal microflora. *American Journal of Clinical Nutrition* 27, 1409-1416.

Freter, R. and Jones, G.W. (1976). Adhesive properties of *Vibrio cholerae:* nature of the interaction with intact mucosal surfaces. *Infection and Immunity* 14, 246-256.

Friis, R.R. (1972). Interaction of L cells and *Chlamydia psittaci:* entry of the parasite and host responses to its development. *Journal of Bacteriology* 110, 706-721.

Frost, A.J. (1975). Selective adhesion of microorganisms to the ductular epithelium of the bovine mammary gland. *Infection and Immunity* 12, 1154-1156.

Frost, A.J., Wanasinghe, D.D. and Woolcock, J.B. (1977). Some factors affecting selective adherence of microorgan-

isms in the bovine mammary gland. *Infection and Immunity* 15, 245-253.
Gabridge, M.G., Barden-Stahl, Y.D., Polisky, R.B. and Engelhardt, J.A. (1977). Differences in the attachment of *Mycoplasma pneumoniae* cells and membranes to tracheal epithelium. *Infection and Immunity* 16, 766-772.
Gangarosa, E.J. and Merson, M.H. (1977). Epidemiological assessment of the relevance of the so-called enteropathogenic serogroups of *Escherichia coli* in diarrhoea. *New England Journal of Medicine* 296, 1210-1213.
Gesner, B. and Thomas, L. (1965). Sialic acid binding sites: role in haemagglutination by *Mycoplasma gallisepticum*. *Science, New York* 151, 590-591.
Gibbons, R.A. (1975). Attachment of oral streptococci to mucosal surfaces. In *Microbiology - 1975*, pp. 127-131. Edited by D. Schlessinger, Washington : American Society for Microbiology.
Gibbons, R.A., Jones, G.W. and Sellwood, R. (1975). An attempt to identify the intestinal receptor for the K88 adhesin by means of a haemagglutination inhibition test using glycoproteins and fractions from sow colostrum. *Journal of General Microbiology* 86, 228-240.
Glynn, A.A., Brumfitt, W. and Howard, C.K. (1971). K-antigens of *Escherichia coli* and renal involvement in urinary-tract infections. *The Lancet* I, 514.
Gorbach, S.L., Kean, B.H., Evans, D.G., Evans, D.J., Jr. and Bessudo, D. (1975). Travellers' diarrhoea and toxigenic *Escherichic coli*. *New England Journal of Medicine* 292, 933-936.
Gorski, F. and Bredt, W. (1977). Studies on the adherence mechanism of *Mycoplasma pneumoniae*. *FEMS Microbiology Letters* 1, 265-267.
Grayston, J.T. and Wang, S.-P. (1975). New knowledge of chlamydiae and the diseases they cause. *Journal of Infectious Diseases* 132, 87-105.
Guentzel, M.N. and Berry, L.J. (1975). Motility as a virulence factor for *Vibrio cholerae*. *Infection and Immunity* 11, 890-897.
Guinee, P.A.M., Jansen, W.H. and Agterberg, C.M. (1976). Detection of the K99 antigen by means of agglutination and immunoelectrophoresis in *Escherichia coli* isolates from calves and its correlation with enterotoxigenicity. *Infection and Immunity* 13, 1369-1377.
Heckels, J.E., Blackett, B., Everson, J.S. and Ward, M.E. (1976). The influence of surface charge on the attachment of *Neisseria gonorrhoeae* to human cells. *Journal of General Microbiology* 96, 359-364.
Hentges, D.J. (1975). Resistance of the indigenous intestinal

flora to the establishment of invading microbial populations. In *Microbiology - 1975*, pp. 116-119. Edited by D. Schlessinger. Washington : American Society for Microbiology.

Hobson, D. and Holmes, K.K. (1977). *Nongonococcal Urethritis and Related Infections*. Washington, DC. : American Society for Microbiology.

Hodgkiss, W., Short, J.A. and Walker, P.D. (1976). Bacterial surface structures. In *Microbial Ultrastructure: The Use of the Electron Microscope*, pp.49-71. Edited by R. Fuller, and D.W. Lovelock, London : Academic Press.

Hohmann, A. and Wilson, M.R. (1975). Adherence of enteropathogenic *Escherichia coli* to intestinal epithelium in vivo. *Infection and Immunity* 12, 866-880.

Hollingdale, M.R. and Manchee, R.J. (1972). The role of mycoplasma membrane proteins in the absorption of animal cells to *Mycoplasma hominis* colonies. *Journal of General Microbiology* 70, 391-393.

Holmgren, J. and Svennerholm, A.-M. (1977). Mechanisms of disease and immunity in cholera: A review. *Journal of Infectious Diseases* 136, Supplement, S105-S112.

Honda, E. and Yanagawa, R. (1974). Agglutination of trypsinised sheep erythrocytes by the pili of *Corynebacterium renale*. *Infection and Immunity* 10, 1426-1432.

Isaacson, R.E. (1977). K99 surface antigen of *Escherichia coli* : purification and partial characterisation. *Infection and Immunity* 15, 272-279.

Isaacson, R.E., Nagy, B. and Moon, H.W. (1977). Colonization of porcine small intestine by *Escherichia coli* : colonization and adhesion factors of pig enteropathogens that lack K88. *Journal of Infectious Diseases* 135, 531-539.

James-Holmquest, A.N., Swanson, J., Buchanan, T.M., Wende, R. D. and Williams, R.P. (1974). Differential attachment by piliated and nonpiliated *Neisseria gonorrhoeae* to human sperm. *Infection and Immunity* 9, 897-902.

Jephcott, A.E., Reyn, A. and Birch-Andersen, A. (1971). *Neisseria gonorrhoeae*. III. Demonstration of presumed appendages to cells from different colony types. *Acta Pathologica et Microbiologica Scandinavica Section B*, 79, 437-439.

Johansson, K.-E. and Hjerten, S. (1974). Localisation of the Tween 20-soluble membrane proteins of *Acholeplasma laidlawii* by crossed immunoelectrophoresis. *Journal of Molecular Biology* 86, 341-348.

Johnston, K.H., Holmes, K.K. and Gotschlich, E.C. (1976). The serological classification of *Neisseria gonorrhoeae*. I. Isolation of the outer membrane complex responsible for serotypic specificity. *Journal of Experimental Medicine* 143, 741-758.

Jones, G.W., Abrams, G.D. and Freter, R. (1976). Adhesive properties of *Vibrio cholerae* : Adhesion to isolated rabbit brush border membranes and haemagglutinating activity. *Infection and Immunity* 14, 232-239.

Jones, G.W. and Freter, R. (1976). Adhesive properties of *Vibrio cholerae* : Nature of the interaction with isolated rabbit brush border membranes and human erythrocytes. *Infection and Immunity* 14, 240-245.

Jones, G.W. and Rutter, J.M. (1972). Role of K88 antigen in the pathogenesis of neonatal diarrhoea caused by *Escherichia coli* in piglets. *Infection and Immunity* 6, 918-927.

Jones, G.W. and Rutter, J.M. (1974). The association of K88 antigen with haemagglutinating activity in porcine strains of *Escherichia coli*. *Journal of General Microbiology* 84, 135-144.

Kaijser, B., Hanson, L.A., Jodal, U., Lidin-Janson, G. and Robbins, J.B. (1977). Frequency of *E. coli* K antigens in urinary-tract infections in children. *Lancet* I, 663-664.

Kellogg, D.S., Jr., Peacock, W.L., Jr., Deacon, W.E., Brown, L. and Pirkle, C.I. (1963). *Neisseria gonorrhoeae*. I. Virulence genetically linked to clonal variation. *Journal of Bacteriology* 85, 1274-1279.

Kuo, C.C. and Grayston, J.T. (1976). Interaction of *Chlamydia trachomatis* organisms and Hela 229 cells. *Infection and Immunity* 13, 1103-1109.

Kuo, C.C., Wang, S.P. and Grayston, J.T. (1972). Differentiation of TRIC and LGV organisms based on enhancement of infectivity by DEAE-dextran in cell culture. *Journal of Infectious Diseases* 125, 313-317.

Kuo, C.C., Wang, S.P. and Grayston, J.T. (1973). Effect of polycations, polyanions and neuraminidase on the infectivity of trachoma-inclusion conjunctivitis and lymphogranuloma venereum organisms in HeLa cells: sialic acid residues as possible receptors for trachoma inclusion conjunctivitis. *Infection and Immunity* 8, 74-79.

Lankford, C.E. and Legsomburana, U. (1965). Virulence factors of choleragenic vibrios. In *Proceedings of the Cholera Research Symposium*. US Public Health Service Publication No. 1328. pp.109-121. Washington, DC : Government Printing Office.

Manchee, R.J. and Taylor-Robinson, D. (1969). Studies on the nature of receptors involved in the attachment of tissue culture cells to mycoplasmas. *British Journal of Experimental Pathology* 50, 66-75.

Meager, A. and Hughes, R.C. (1977). Virus receptors. In *Receptors and Recognition*, Series A, vol. 4, pp. 142-195. Edited by P. Cuatrecases and M.F. Greaves. London : Chapman and Hall.

McEntegart, M.G. (1975). Gonorrhoea : Fundamental research developments. In *Recent Advances in Sexually Transmitted Diseases*, pp. 54-68. Edited by R.S. Morton and J.R.W. Harris. Edinburgh : Churchill Livingstone.

McNeish, A.S., Turner, P., Fleming, J. and Evans, N. (1975). Mucosal adherence of human enteropathogenic *Escherichia coli*. *Lancet* II, 946-948.

Mims, C. (1976). *Pathogenicity of Infectious Disease*. London and New York : Academic Press.

Moon, H.W., Nagy, B. and Isaacson, R.E. (1977). Intestinal colonization and adhesion by enterotoxigenic *Escherichia coli:* ultrastructural observations on adherence to ileal epithelium in the pig. *Journal of Infectious Diseases* 136, Supplement, S124-S129.

Nagy, B., Moon, H.W. and Isaacson, R.E. (1976). Colonization of porcine small intestine by *Escherichia coli:* ileal colonization and adhesion by pig enteropathogens that lack K88 antigen and by some acapsular mutants. *Infection and Immunity* 13, 1214-1220.

Nagy, B., Moon, H.W. and Isaacson, R.E. (1977). Colonization of porcine intestine by enterotoxigenic *Escherichia coli:* Selection of piliated forms in vivo, adhesion of piliated forms to epithelial cells in vitro, and incidence of a pilus antigen among porcine enteropathogenic *E. coli*. *Infection and Immunity* 16, 344-352.

Nelson, E.T., Clements, J.D. and Finkelstein, R.A. (1976). *Vibrio cholerae* adherence and colonization in experimental cholera: Electron microscopic studies. *Infection and Immunity* 14, 527-547.

Novotny, P., Short, J.A., Hughes, M., Miler, J.J., Syrett, C., Turner, W.H., Harris, J.R.W. and MacLennon, I.P.B. (1977). Studies on the mechanism of pathogenicity of *Neisseria gonorrhoeae*. *Journal of Medical Microbiology* 10, 347-366.

Novotny, P., Short, J.A. and Walker, P.D. (1975). An electron microscope study of naturally occurring and cultured cells of *Neisseria gonorrhoeae*. *Journal of Medical Microbiology* 8, 413-427.

Novotny, P. and Turner, W.H. (1975). Immunological heterogeneity of pili of *Neisseria gonorrhoeae*. *Journal of General Microbiology* 89, 87-92.

Ofek, I., Beachy, E.H., Jefferson, W. and Campbell, G.L. (1965). Cell membrane-binding properties of Group A streptococcal lipoteichoic acid. *Journal of Experimental Medicine* 141, 990-1003.

Ofek, I., Mirelman, D. and Sharon, N. (1977). Adherence of *Escherichia coli* to human mucosal cells mediated by mannose receptors. *Nature, London* 265, 623-625.

Old, D.C. (1972). Inhibition of the interaction between fimbrial haemagglutinins and erythrocytes by D-mannose and other carbohydrates. *Journal of General Microbiology* 71, 149-157.

Øsrkov, F., Ørskov, I., Evans, D.J., Jr., Sack, D.A. and Wadstrom, T. (1976). Special *Escherichia coli* serotypes among enterotoxigenic strains from diarrhoea in adults and children. *Medical Microbiology and Immunology* 162, 73-80.

Ørskov, I. and Ørskov, F. (1966). Episome-carried surface antigen K88 of *Escherichia coli*. I. Transmission of the determinant of the K88 antigen and the influence on the transfer of chromosomal markers. *Journal of Bacteriology* 91, 69-75.

Ørskov, I. and Ørskov, F. (1977). Special O:K:H serotypes among enterotoxigenic *E. coli* strains from diarrhoea in adults and children. Occurrence of the CF (colonization factor) antigen and of haemagglutinating activities. *Medical Microbiology and Immunology* 163, 99-110.

Ørskov, I., Ørskov, F., Smith, H. Williams- and Sojka, W.J. (1975). Establishment of K99, thermolabile, *Escherichia coli* K antigen, previously called "Kco" possessed by calf and lamb enteropathogenic strains. *Acta Pathologica et Microbiologica Scandanavica Section B* 83, 31-36.

Pedersen, K.B., Froholm, L.O. and Bovre, K. (1972). Fimbriation and colony type of *Moraxella bovis* in relation to conjunctival colonization and development of keratoconjunctivitis in cattle. *Acta Pathologica et Microbiologica Scandinavica Section B* 80, 911-918.

Penn, C.W., Parsons, N.J., Sen, D., Veale, D.R. and Smith, H. (1977). Immunization of guinea pigs with *Neisseria gonorrhoeae:* Strains specificity and mechanisms of immunity. *Journal of General Microbiology* 100, 159-166.

Penn, C.W., Sen, D., Veale, D.R., Parsons, N.J. and Smith, H. (1976). Morphological biological and antigenic properties of *Neisseria gonorrhoeae* adapted to growth in guinea-pig subcutaneous chambers. *Journal of General Microbiology* 97, 35-43.

Penn, C.W., Veale, D.R. and Smith, H. (1977). Selection from gonococci grown *in vitro* of a colony type with some virulence properties of organisms adapted in vivo. *Journal of General Microbiology* 100, 147-158.

Powell, D.A., Hu, P.C., Wilson, M., Collier, A.M. and Baseman, J.B. (1976). Attachment of *Mycoplasma pneumoniae* to respiratory epithelium. *Infection and Immunity* 13, 959-966.

Pugh, G.W. and Hughes, D.E. (1976). Experimental production of infectious bovine keratoconjunctivitis: comparison of serological and immunological responses using pili fractions of *Moraxella bovis*. *Canadian Journal of Comparative*

Medicine 40, 60-66.
Reimann, R. and Lind, I. (1977). An indirect haemagglutination test for demonstration of gonococcal antibodies using gonococcal pili as antigen. *Acta Pathologica et Microbiologica Scandinavica Section C* 85, 115-122.
Robertson, J.N., Vincent, P. and Ward, M.E. (1977). The preparation and properties of gonococcal pili. *Journal of General Microbiology* 102, 169-177.
Rowland, M.G.M. and McCollum, J.P.K. (1977). Malnutrition and gastroenteritis in The Gambia. *Transactions of the Royal Society of Tropical Medicine and Hygiene* 71, 199-203.
Rutter, J.M., Jones, G.W., Brown, G.T.H., Burrows, M.R. and Luther, P.D. (1976). Antibacterial activity in colostrum and milk associated with protection of piglets against enteric disease caused by K88-positive *Escherichia coli*. *Infection and Immunity* 13, 667-676.
Sack, R.B., Gorbach, S.L., Banwell, J.G., Jacobs, B., Chatterjee, B.D. and Mitra, R.C. (1971). Enterotoxigenic *Escherichia coli* isolated from patients with severe cholera-like disease. *Journal of Infectious Diseases* 123, 378-385.
Sack, D.A., Merson, H.M., Wells, J.G., Sack, R.B. and Morris, G.K. (1975). Diarrhoea associated with heat-stable enterotoxin-producing strains of *Escherichia coli*. *Lancet* II, 239-241.
Salit, I.E. and Gotschlich, E.C. (1976). Characterisation of *E. coli* binding sites on mammalian cell membranes. *Abstracts of Interscience Conference on Antimicrobial Agents and Chemotherapy* 16, No. 10.
Savage, D.C. (1972). Associations and physiological interactions of indigenous microorganisms and gastrointestinal epithelia. *American Journal of Clinical Nutrition* 25, 1372-1379.
Savage, D.C. (1975). Indigenous microorganisms associating with mucosal epithelia in the gastrointestinal ecosystem. In *Microbiology - 1975*, pp. 120-123. Edited by D. Schlessinger. Washington : American Society for Microbiology.
Savage, D.C. (1976). Microbial colonisation of epithelial surfaces in the intestinal tract. In *Proceedings 'Microbial Aspects of Dental Caries'*. Special supplement Microbiology Abstracts, pp. 33-46. Edited by H.M. Stiles, W.J. Loesche and T.C. O'Brien. Washington and London : Information Retrieval Inc.
Savage, D.C. and Blumershine, R.V.H. (1974). Surface-surface associations in microbial communities populating epithelial habitats in the murine gastrointestinal ecosystem: scanning electron microscopy. *Infection and Immunity* 10, 240-250.

Schrank, G.D. and Verwey, W.F. (1976). Distribution of cholera organisms in experimental *Vibrio cholerae* infections: Proposed mechanisms of pathogenesis and antibacterial immunity. *Infection and Immunity* 13, 195-203.

Sellwood, R., Gibbons, R.A., Jones, G.W. and Rutter, J.M. (1975). Adhesion of enteropathogenic *Escherichia coli* to pig intestinal brush borders : the existence of two pig phenotypes. *Journal of Medical Microbiology* 8, 405-411.

Shinefield, H.R., Ribble, J.C., Boris, M. and Eichenwald, H.F. (1972). Bacterial interference. In *The Staphylococci*, pp. 503-515. Edited by J.O. Cohen. New York and London : John Wiley and Sons Inc.

Shore, E.G., Dean. A.G.. Holik, K.J. and Davis, B.R. (1974). Enterotoxin producing *Escherichia coli* and diarrhoeal disease in human travelers: a prospective study. *Journal of Infectious Diseases* 129, 577-582.

Silverblatt, F.J. and Ofek, I. (1976). Effect of pili on virulence of *P. mirabilis* in experimental pyelonephritis. *Abstracts of Interscience Conference on Antimicrobial Agents and Chemotherapy* 16, No.12.

Smith, H. (1977). Microbial surfaces in relation to pathogenicity. *Bacteriological Reviews* 41, 475-500.

Smith, H. Williams- and Linggood, M.A. (1971). Observations on the pathogenic properties of the K88, HLY, and ENT plasmids of *Escherichia coli* with particular reference to porcine diarrhoea. *Journal of Medical Microbiology* 4, 467-485.

Smith, H. Williams- and Linggood, M.A. (1972). Further obser vations on *Escherichia coli* enterotoxins with particular regard to those produced by atypical piglet strains and by calf and lamb strains; the transmissible nature of these enterotoxins and of a K antigen possessed by calf and lamb strains. *Journal of Medical Microbiology* 5, 243-250.

Smyth, C.J. and Salton, M.R.J. (1977). Crossed immunoelectrophoresis. A new approach to high resolution analysis of gonococcal antigens and antibodies. In *The Gonococcus*, pp. 303-332. Edited by R.B. Roberts. New York : John Wiley and Sons Inc.

Smyth, C.J., Siegel, J., Salton, M.R.J. and Owen, P. (1978). Immunochemical analysis of inner and outer membranes of *Escherichia coli* by crossed immunoelectrophoresis. *Journal of Bacteriology* 133, 306-319.

Sobeslavsky, O., Prescott, B. and Chanock, R.M. (1968). Adsorption of *Mycoplasma pneumoniae* to neuraminic acid receptors of various cells and possible role in virulence. *Journal of Bacteriology* 96, 695-705.

Sojka, W.J. (1965). *Escherichia coli* in domestic animals and poultry. *Review Series No.7 Commonwealth Bureau of Animal*

Health, Farnham Royal.

Stirm, S., Ørskov, F., Ørskov, I. and Mansa, B. (1967a). Episome-carried surface antigen K88 of *Escherichia coli*. Isolation and chemical analysis. *Journal of Bacteriology* 93, 731-739.

Stirm, S., Ørskov, F., Ørskov, I. and Birch-Andersen, A. (1967b). Episome-carried surface antigen K88 of *Escherichia coli*. Morphology. *Journal of Bacteriology* 93, 740-748.

Swanson, J. (1972). Studies on gonococcus infection. II. Freeze-fracture, freeze-etch studies of gonococci. *Journal of Experimental Medicine* 136, 1258-1271.

Swanson, J. (1977). Surface components affecting interactions between *Neisseria gonorrhoeae* and eucaryotic cells. *Journal of Infectious Diseases* 136, Supplement, S138-S143.

Swanson, J., Kraus, S.J. and Gotschlich, E.C. (1971). Studies on gonococcus infection. 1. Pili and zones of adhesion. Their relation to gonococcal growth patterns. *Journal of Experimental Medicine* 134, 886-906.

Swanson, J., Sparks, E., Zeligs, B., Siam, M.A. and Parrott, C. (1974). Studies on gonococcus infection. V. Observations on *in vitro* interactions of gonococci and human neutrophils. *Infection and Immunity* 10, 633-644.

Tannock, G.W., Blumershine, R.V. and Savage, D.C. (1975). Association of *Salmonella typhimurium* with, and its invasion of, the ileal mucosa in mice. *Infection and Immunity* 11, 365-370.

Tebbutt, G.M., Veale, D.R., Hutchinson, J.G.P. and Smith, H. (1976). The adherence of pilate and non-pilate strains of *Neisseria gonorrhoeae* to human and guinea-pig epithelial tissues. *Journal of Medical Microbiology* 9, 263-273.

Tebbutt, G.M., Veale, D.R., Penn, C.W. and Smith, H. (1977). The adherence multiplication and localisation of *in vitro* and *in vivo* grown gonococci on human mucosal tissues. *FEMS Microbiology Letters* 1, 171-174.

Tramont, E.C. and Wilson, C. (1977). Variations in buccal cell adhesion of *Neisseria gonorrhoeae*. *Infection and Immunity* 16, 709-711.

Wagner, R.C. and Barrnett, R.J. (1974). The fine structure of prokaryotic-eukaryotic cell junctions. *Journal of Ultrastructural Research* 48, 404-413.

Ward, M.E. and Watt, P.J. (1975). Studies on the cell biology of gonorrhoeae. In *Genital Infections and Their Complications*. pp. 229-242. Edited by D. Danielsson, L. Johlin and P.-A, Mardh. Stockholm : Almquist and Wiksell International.

Ward, M.E., Robertson, J.N., Englefield, P.M. and Watt, P.J. (1975). Gonococcal infection: Invasion of the mucosal surfaces of the genital tract. In *Microbiology - 1975* pp.

188-199. Edited by D. Schlessinger. Washington : American Society for Microbiology.

Ward, M.E., Watt, P.J. and Glynn, A.A. (1970). Gonococci in urethral exudates possess a virulence factor lost on subculture. *Nature, London* 227, 382-384.

Watt, P.J. and Ward, M.E. (1977). The interaction of gonococci with human epithelial cells. In *The Gonococcus,* pp. 356-368. Edited by R.B.Roberts. New York : John Wiley and Sons Inc.

Watt, P.J., Ward, M.E. and Robertson, J.N. (1976). Interaction of gonococci with host cells. In *Sexually Transmitted Diseases* pp. 89-101. Edited by R.D. Catterall and C.S. Nicol. London : Academic Press.

Weiss, E. and Dressler, H.R. (1960). Centrifugation of rickettsia and viruses onto cells and its effects on infection. *Proceedings of the Society for Experimental Biology and Medicine* 103, 691-695.

Williams, H.R., Jr., Verwey, W.F., Schrank, G.D. and Hurry, E.K. (1973). An *in vitro* antigen-antibody reaction in relation to a hypothesis of intestinal immunity to cholera. In *Proceedings of the 9th Joint Cholera Research Conference, Washington,* pp. 161-173. US Department of State Publication No. 8762.

Williams, P.H., Evans, N., Turner, P., George, R.H. and McNeish, A.S. (1977). Plasmid mediating mucosal adherence in human enteropathogenic *Escherichia coli*. *Lancet* I, 1151.

SUMMING UP

A.S.G. CURTIS

Department of Cell Biology, University of Glasgow, Glasgow G12 8QQ. Scotland.

Work on biological adhesion has mostly been directed to discovering the mechanism or mechanisms by which eukaryote cells adhere to each other or to extended non-living surfaces. However, many of the chapters in this book will remind the reader of the biological, economic and theoretical importance of studies on the mechanisms by which bacteria and other microorganisms adhere to a variety of materials such as teeth, metal, mineral, ceramic and plastic surfaces, or to cells such as those of our tissues. Thus if work in this field has started relatively late there is the advantage of having the speculations, results and theories of the workers on eukaryote cells to act as a guide.

Work on eukaryote cells has proceeded in a number of rather different directions but several themes are worth examination to discover their possible relevance to microorganism adhesion. These themes can be identified by asking several questions.

What is the magnitude of the various physical forces acting in determining the approach of cells to surfaces, in establishing and maintaining an adhesion, and in preventing reseparation?

The answers to this have been outlined in Chapter 2 (page 5) to a considerable degree and also in reviews by myself [Curtis, 1973; 1978]. The question refers, of course, to experimental attempts to quantify adhesion and thus to identify these forces. In many situations shear forces in either laminar or turbulent flow ensure that the particles make rapid close approaches. In the absence of any adhesive ones, of course, the particles will normally reseparate. For instance if a flow impinges on a plate, particles will be carried onto the surface with a normal direction; the flow will then carry the particles away laterally and radially. The only exception to this is if sedimentation takes place in a centrifugal or gravitational field. If it was possible to identify the magnitude and range of the force required to perturb interactions so that capture occurs under given approach conditions then this would go a long way to identifying the forces that can bring about cell adhesion, and to measuring adhesion quantitatively. Similarly, the time required for formation of an adhesion might be a valuable aid to identifying the forces that ensure that an approach leads to an adhesion.

Unfortunately the colloid scientists are still faced with theoretical problems which at present rather limit these methods. The collision rate for laminar shear flow and Brownian motion collision introduced by von Smoluchowski was refined to cover the integration over all aggregate classes by Swift and Friedlander [1964]. This notable advance allowed the introduction (by Dr. Hocking and myself [Curtis and Hocking, 1970]) of a method of measuring the probability that particles would adhere on collision in a laminar shear flow. In this we took into account the shear-induced forces that bring about collision and reseparation and the drainage forces between the particles that tend to prevent approach and also reseparation. Somewhat similarly Weiss and Harlos [1972] considered Brenner's [1961] treatment of the approach of a sphere to a horizontal surface as a system in which the duration of approach might give information on the extent to which this could be perturbed by forces of attraction or repulsion. Doroszewski (in press) has introduced a general treatment for the adhesion of particles to the wall of a tube through which they are flowing.

Unfortunately these general approaches suffer from a number of possible defects. Firstly, these approaches are modelled on the behaviour of large bodies and ignore such effects as changes in viscosity very close to a surface. Secondly, the particles involved are assumed to be hard undeformable spheres, even though they may be as soft as oil globules. Fortunately for microbiologists who wish to use 'exact' treatments, prokaryotes and other microorganisms are often more akin to hard spheres than are eukaryotic cells.

It has often been argued that the forces that maintain adhesions are different from those that initiate them. Though this may be so, very little concrete evidence on the matter has been obtained. In any event, in practical terms it may be much more important to know how adhesions are established rather than how they are maintained. Such processses as fouling, infection of cells and a variety of treatment procedures require the formation of adhesions and should these be prevented by interfering with the initial steps then it would be possible to control these important processes.

Once an adhesion has been formed, the main quantitative method of investigating it is to attempt to obtain a measure of the force required to break it. This technique has often been used in a very imprecise and subjective manner, though more precisely by Brooks *et al.* [1967] and for non-living systems by Visser [1976] whose methods have considerable elegance and potential application to microbiological research. Once again it is likely that more precise results will be obtained for prokaryote and walled eukaryote cells than for naked eukaryote cells because there will be little problem with the question as to whether the cells detach by peeling or in one single step.

However, it is clear that only very precisely controlled experiments will be of real value, because changes in such things as the medium viscosity, particle motion and shear regimes, produce effects on the forces acting to form or break adhesions which are non-linear and almost impossible to assess intuitively. To a considerable extent data is required for physical studies of ideal systems which should provide a better understanding of the sort of experimental methods that should be used.

The forces that form an adhesion are aided by Brownian motion in that particles the size of microbes will be frequently brought into collision by Brownian motion even if shear induced processes are also important. In this respect they differ from eukaryote cells which are too large for Brownian motion to play much of a role in their total motion. However, it is possible that both the pili and fimbriae on bacterial cells and the microvilli on eukaryote cells may be brought into interaction by Brownian motion. Pethica [1961] was the first person to make this suggestion but unfortunately his model is probably oversimplified in that he assumed that such fine protrusions could be treated like independent spheres moved by thermal motion unattached to their cells. In practice there is little evidence that microvilli aid adhesion in eukaryotes and some evidence that pili may actually be associated with a reduction in adhesion of *Neisseria* to macrophages [Swanson, 1977].

There are several other ways in which the effects of Brownian motion on microbial adhesion can be examined. First, the probability that an adhesion forms should be a direct consequence of the competition, or alternatively synergism, between the forces of adhesion acting on an approaching particle and on the probability that the particle receives a second Brownian motion hit with energy sufficient to drive it away (competition) or further towards the surface (synergism) during the period that the particle is in range for the forces of adhesion to act. If the probability of a distracting second hit is high within the period before which the forces of attraction are able to bring the particles into a stable adhesion, then adhesions will form only rarely. Examination of the spectrum of possible Browian motion energies acting on a particle and on the observed probability of an adhesion forming might give information on the range and nature of forces of attraction that must act to overcome Brownian motion inhibition of adhesion.

Dr. Doowy of the Institute of Mathematical Biophysics, Warsaw, has also pointed out that Brownian motion may have another effect on adhesion if it is essential for the two adhering bodies to be totally immobile relative to each other during the formation of an adhesion. If this condition is to be satisfied Brownian motion may have to act temporarily to prevent any shear induced motion during the formation of an adhesion. Again this probability can be evaluated for various energies of approach. Clearly it should be worthwhile directing thought towards the relevance of these factors.

Which forces are important in adhesion of cells?

Unfortunately there are considerable difficulties in proving that either a DLVO system, or any other of the possible models that are provided by colloid chemistry, acts in a given system and it may be possible to interpret every piece of evidence that has been claimed to support a particular mechanism in an alternative manner. My own feeling is that really rigorous examination of the kinetics of formation of adhesions may provide more precise information than any other single approach, combined perhaps with a careful study of the surface chemistry involved and of the effects of modification of this chemistry.

Modification of the cell surface lipids has been attempted by myself and co-workers [Curtis *et al.*, 1975; Schaeffer and Curtis, 1977] using the re-acylation system of the plasmalemma and by inclusion of liposomes [Pagano and Takeichi, 1977] with extensive effects on cell adhesion. The results show that modification has to be extensive in order to have marked effects on cell adhesion, which might argue that some generalised surface property such as electrodynamic forces is involved in adhesion. The next stage in the argument however reveals the present ambiguities of such an approach. One argument which follows from that conclusion is that adhesion is due to an averaged feature of cell surfaces involving relatively large areas of surface. However, a second argument is that the effects may be indirect: for instance, cell surface lipids may affect fluidity and in turn this may affect the aggregation of some quite rare species of molecule, required to be in groups for adhesion. Alternatively it might be supposed that the state of the cell surface lipids might affect the orientation and/or adsorption of, for example, surface glycoproteins.

Temperature effects may represent perturbations in enzyme systems directly or indirectly involved in adhesion, but also can be interpreted as being effects on lipid or glycolipid structure either acting on fluidity or on electrodynamic forces.

Though it is clear from circumstantial evidence [see Curtis, 1973] that electrostatic forces of repulsion probably play a large, though not total, role in controlling eukaryote adhesion the answers may be rather different for microbial adhesion. Some microbes can live in media of such high ionic strength that electrostatic forces of repulsion must be very short range, while other microbes live in such low ionic strength media that these forces may be of permanent importance, affecting interactions at a very long range.

A series of questions needs answering experimentally before the role of electrostatic forces in adhesion can really be evaluated. Amongst these questions are (i) do cell surfaces or microbial walls carry polarisable electrical double layers whose range would be affected by electrogenesis by the cells involved? - (ii) do cations, such as Fe(III) absorb to biological surfaces sufficiently to affect electrostatic interactions even though their bulk concentrations may be very low? Nordin *et al.* [1967] showed that Fe(III) affects the adhesion of *Chlorella* at very low levels. Weinberg [1974] reviewed the role of iron in microbe-eukaryote host interactions. Are these interactions primarily adhesive ones?

Is there clear evidence that identifiable chemical groupings are directly involved in adhesion forming ionic or covalent bondings between cells?

Two approaches have been made to answer this question. In the first, a variety of reagents have been used to treat cells, and the cells then have been tested to discover if adhesion is changed. Such treatments include the use of enzymes of fairly high specificity which should cleave (or occasionally add) a precise grouping. For instance trypsin modifies the adhesion of eukaryote cells both to themselves and to non-cellular surfaces. Does this argue that a trypsin sensitive protein or peptide is directly involved in adhesion?

Unfortunately this result does not have such a clear interpretation, even if it is possible to establish that trypsin is acting only at the cell surface. The reason for this is that trypsin affects the enzymes involved in turnover of lipids and allows the accumulation of lysophospholipids in the plasmalemma [Curtis in preparation]. Lysophospholipids affect cell adhesion either directly or indirectly, [Curtis *et al.*, 1975]. Thus the effects of trypsin are at least in part very indirect.

The second approach is in many ways more subtle, and starts with the proposition that mutants can be isolated, which show changed adhesive properties. If the expression of the mutation can then be tied down to a precise biochemical change at the cell surface it is tempting to conclude that this must play some very important role in adhesion. For instance Swanson has shown that the adhesion of *Neisseria gonorrhoeae* to leucocytes is determined by a protein that is found only in strains that show this phenomenon [Swanson, 1977]. Again the K88 and K99 antigens found in certain strains of *Escherichia coli* are associated with pathogenicity in pigs and calves [Jones and Rutter, 1974] and may well be involved in adhesion to intestinal mucosa. However, these experiments do not provide proof that the antigens or proteins are directly involved in adhesion.

Ofek *et al.* [1977] showed that the adhesion of *Escherichia coli* to epithelial human mucosa cells could be inhibited by D-mannose and methyl D-mannopyranoside, and that a mannose binding protein could be extracted from the bacteria. They conclude that mannose, found in many vertebrate plasmalemmal glycoproteins, acts as a binding receptor for this bacterium. Though this conclusion may be correct I would suggest that it is premature to come to such a deduction.

It is also worth considering the various roles that macromolecules adsorbed from the medium may play in adhesion. Adsorption may tend to prevent adhesion by stopping close approach of the surfaces or by increasing electrostatic forces of repulsion, or aid adhesion by reducing electrostatic forces of repulsion, or by entanglement between adsorbed layers on two surfaces. It is important to bear in mind that enzyme secretion by cells may tend to lyse such adsorbed materials and that eukaryote cells possess 'cleaning' mechanisms which may remove adsorbed materials by phagocytosis.

Table 1 lists the points that must be experimentally proven before it can be concluded that a specific grouping is involved in adhesion. No report yet made, as far as I am aware, meets all of them. Until these points are covered it is not clear whether the binding systems act directly in adhesion or indirectly.

It is questionable whether such binding systems could perturb an interaction sufficiently to ensure capture without the action of some relatively long range attraction. Can particles the size of microbes make approach closely enough and rapidly enough for covalent, hydrogen or ionic bonding before Brownian motion hits tend to distract the particles? Clearly centrifugation under a sufficient force will effect such an approach.

Table 1.

Requirements for the demonstration that a given surface component is directly involved in adhesion by forming a complex with another cell surface or with an intermediate bridging molecule

1. Isolation and purification of the molecular species involved.
2. Demonstration of the surface location of this molecule
3. Cells denuded of the molecule should either not show adhesion or have a much reduced adhesion.
4. Cells re-coated with a normal level of the molecule should re-acquire their original adhesive properties.
5. Dose response curve should be of a form appropriate to the valency number of the molecules and receptors involved.
6. Contacts formed should be of the 'bridging type'.
7. Antibody (Fab) to the molecule may inhibit adhesion.
8. Enzymic dissection of the molecule should probably destroy adhesion.

But Brownian motion induced collision and collision induced by shear may not provide sufficient interaction unless aided by a long range force such as an electrodynamic one.

Does the specificity of adhesion seen in many situations argue that very precise molecular mechanisms are involved?

Sponge cells show a species-specific aggregation and many vertebrate embryonic cells show tissue specificity in their preferred adhesion [see review by Curtis, 1978]. These phenomena have suggested to many that adhesion must involve very specific biochemical mechanisms even though it has been pointed out on several occasions that specific adhesion cannot explain the precise patterning found in the segregation of embryonic cells [Curtis, 1962; Steinberg, 1962]. Thus it seemed that the specificity of adhesion might be more complex than appears at first sight. I and Gysele van de Vyver [Curtis and van de Vyver, 1971] have shown that different strains of a sponge species produce soluble diffusible factors that diminish the adhesion of unlike cell types. Subsequently [Curtis and De Sousa, 1975; Curtis, 1978] it was shown that many cell types produce small proteins, termed interaction modulation factors, that diminish cell adhesion in their target cells.

Table 2.

Critical paths to adhesion

(Particle, eg. microbe, to large cell or non-living surface)

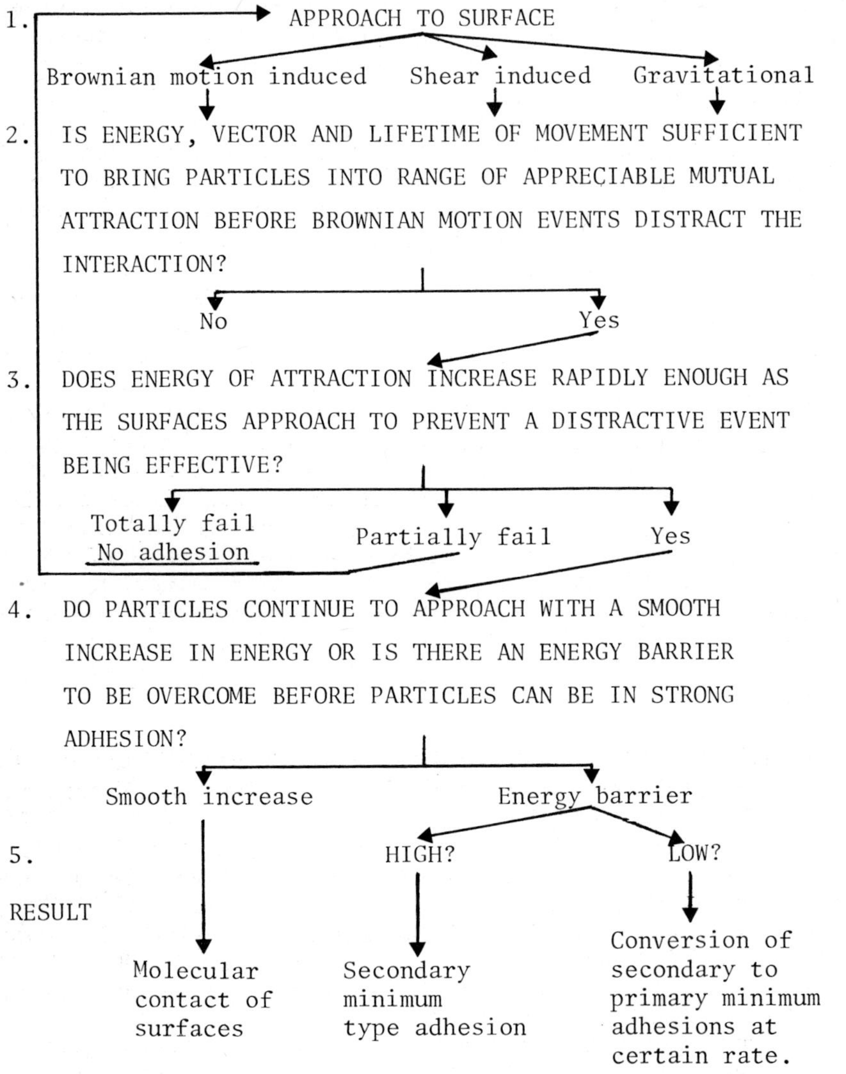
1.
APPROACH TO SURFACE
Brownian motion induced
Shear induced
Gravitational
2.
IS ENERGY, VECTOR AND LIFETIME OF MOVEMENT SUFFICIENT TO BRING PARTICLES INTO RANGE OF APPRECIABLE MUTUAL ATTRACTION BEFORE BROWNIAN MOTION EVENTS DISTRACT THE INTERACTION?
No
Yes
3.
DOES ENERGY OF ATTRACTION INCREASE RAPIDLY ENOUGH AS THE SURFACES APPROACH TO PREVENT A DISTRACTIVE EVENT BEING EFFECTIVE?
Totally fail
No adhesion
Partially fail
Yes
4.
DO PARTICLES CONTINUE TO APPROACH WITH A SMOOTH INCREASE IN ENERGY OR IS THERE AN ENERGY BARRIER TO BE OVERCOME BEFORE PARTICLES CAN BE IN STRONG ADHESION?
Smooth increase
Energy barrier
5.
HIGH?
LOW?
RESULT
Molecular contact of surfaces
Secondary minimum type adhesion
Conversion of secondary to primary minimum adhesions at certain rate.

Such proteins produce a specific control of adhesion, not by stimulating adhesion amongst the producer cells but by diminishing the adhesion of unlike cells. Since the factors are diffusible it is likely that concentration gradients of these substances will be set up and that these may produce patterning of cells. It is interesting to speculate as to whether similar mechanisms may operate in interactions between microorganisms or between microorganisms and eukaryote cells. For instance, do eukaryote cells use such mechanisms to prevent the attachment of pathogenic bacteria?

My purpose in writing this summing-up is not so much to provide a review of those results which have gone before but rather to indicate the next round of problems and concepts that we shall have to face. In conclusion it must be remembered that microbial adhesion occurs in a very wide variety of conditions, for example, ionic strengths ranging from nearly pure water to saline solutions in excess of 3M-NaCl, pH from low to fairly high values, temperatures from supercooled to nearly boiling water, with adhesion sometimes in media of relatively low or high dielectric constant and with many unusual chemicals present on occasion. Eukaryote adhesion, at least of live eukaryote cells, occurs in a much more restricted range of conditions. Thus it is unlikely that a single mechanism can explain microbial adhesion. However it is likely that the formation of an adhesion goes through the flow path indicated by Table 2. Quantitative evaluation of each of these stages for a given system should allow a very precise answer to be provided for the question of how does such-and-such a microbe adhere and how does it maintain its adhesion.

REFERENCES

Brenner, H. (1961). The slow motion of a sphere through a viscous fluid towards a plane surface. *Chemical Engineering Sience* 16, 242-251.

Brooks, D.E., Millar, J.S., Seaman, G.V.E. and Vassar, P.S. (1967). Some physiochemical factors relevant to cellular interactions. *Journal of Cellular Physiology* 69, 155-168.

Curtis, A.S.G. (1962). Cell contact and adhesion. *Biological Reviews of the Cambridge Philosophical Society* 37, 82-129.

Curtis, A.S.G. (1973). Cell Adhesion. *Progress in Biophysics and Molecular Biology* 27, 315-386.

Curtis, A.S.G. (1978). Cell positioning. To be published in *Specificity in Embryological Interactions*. Edited by D. Garrod *Receptors and Recognition* Series B, vol. 2.

Curtis, A.S.G., Chandler, C. and Picton, N. (1975). Cell surface lipids and adhesion. III. The effects on cell adhesion of changes in plasmalemmal lipids. *Journal of Cell*

Science 18, 375-384.
Curtis, A.S.G. and De Sousa, M. (1975). Lymphocyte interactions and positioning. *Cellular Immunology* 19, 282-297.
Curtis, A.S.G. and Hocking, L.M. (1970). Collision efficiency of equal spherical particles in shear flow. The influence of the London-van der Waals forces. *Transactions of the Faraday Society* 66, 1381-1390.
Curtis, A.S.G. and Van de Vyver, G. (1971). The control of cell adhesion in a morphogenetic system. *Journal of Embryology and Experimental Morphology* 26, 295-312.
Jones, G.W. and Rutter, J.M. (1974). The association of K88 antigen with haemagglutinating activity in porcine strains of *Escherichia coli*. *Journal of General Microbiology* 84, 135-144.
Nordin, J.S., Tsuchiya, H.M. and Fredrickson, A.G. (1967). Interfacial phenomena governing adhesion of Chlorella to glass surfaces. *Biotechnology and Bioengineering* 9, 545-558.
Ofek, I., Mirelman, D. and Sharon, N. (1977). Adherence of *Escherichia coli* to human mucosal cells mediated by mannose receptors. *Nature, London* 265, 623-625.
Pagano, R.E. and Takeichi, M. (1977). Adhesion of phospholipid vesicles to chinese hamster fibroblasts. *Journal of Cell Biology* 74, 531-546.
Pethica, B.A. (1961). The physical chemistry of cell adhesion. *Experimental Cell Research* Suppl 8, 123-140.
Schaeffer, B.E. and Curtis, A.S.G. (1977). Effects on cell adhesion and membrane fluidity of changes on plasmalemmal lipids in mouse L929 cells. *Journal of Cell Science* 26, 47-55.
Steinberg, M.S. (1963). On the mechanism of tissue reconstruction by dissociated cells. II. Time course of events. *Science* 137, 762-763.
Swanson, J. (1977). Surface components affecting interactions between *Neisseria gonorrhoeae* and eucaryotic cells. *Journal of Infectious Diseases* 136, S138-S143.
Swift, D.L. and Friedlander, S.K. (1964). The coagulation of hydrosols by Brownian motion and laminar shear flow. *Journal of Colloid Science* 19, 621-647.
Visser, J. (1976). The adhesion of colloidal polystyrene particles to cellophane as a function of pH and ionic strength. *Journal of Colloid and Interface Science* 55, 664-677.
Weinberg, E.D. (1974). Iron and susceptibility to infectious disease. *Science* 184, 952-956.
Weiss, L. and Harlos, J.P. (1972). Some speculations of the rate of adhesion of cells to coverslips. *Journal of Theoretical Biology* 37, 169-179.

SUBJECT INDEX